FOR THE
IB DIPLOMA

Internal Assessment for
Physics

Christopher Talbot
Ged Green

Skills for Success

Authors' acknowledgements

We thank Mr Cesar Reyes and Dr David Fairley of the Overseas Family School, Singapore, who reviewed and made many useful suggestions on early drafts of the material. We are also grateful to the latter for his contribution of Python programs and accompanying notes specially written for this publication. We also thank John Allum (author of *Physics for the IB Diploma* and of *Physics for the IB Diploma Study and Revision Guide*) for his detailed review of the entire manuscript and his many helpful suggestions and comments, in addition to proofreading all of the chapters. Adam Hicks, Wellington College, gave helpful feedback on the chapters centred on assessment criteria.

Photo credits

The Publishers would like to thank the following for permission to reproduce copyright material.

p. 13, Figure 2.14: photo courtesy of Philip Harris. p. 16, Figure 2.18: Edulab, Figure 2.19: Edulab. p. 23, Figure 2.34: Edulab. p. 24, Figure 2.35: photo courtesy of Philip Harris. p. 33, Figure 3.11: photo courtesy of Philip Harris. pp. 80–84, Figures 6.1, 6.2, 6.3, 6.4 and 6.5: Dr David Fairley.

Every effort has been made to trace all copyright holders, but if any have been inadvertently overlooked, the Publishers will be pleased to make the necessary arrangements at the first opportunity.

Although every effort has been made to ensure that website addresses are correct at time of going to press, Hodder Education cannot be held responsible for the content of any website mentioned in this book. It is sometimes possible to find a relocated web page by typing in the address of the home page for a website in the URL window of your browser.

Hachette UK's policy is to use papers that are natural, renewable and recyclable products and made from wood grown in well-managed forests and other controlled sources. The logging and manufacturing processes are expected to conform to the environmental regulations of the country of origin.

Orders: please contact Hachette UK Distribution, Hely Hutchinson Centre, Milton Road, Didcot, Oxfordshire, OX11 7HH. Telephone: +44 (0)1235 827827. Email: education@hachette.co.uk. Lines are open from 9 a.m. to 5 p.m., Monday to Friday. You can also order through our website: www.hoddereducation.com

ISBN: 978 1 5104 3241 3

© Christopher Talbot and Ged Green 2019

First published in 2019 by

Hodder Education (a trading division of Hodder & Stoughton Limited),

An Hachette UK Company

Carmelite House

50 Victoria Embankment

London EC4Y 0DZ

www.hoddereducation.com

The authorised representative in the EEA is Hachette Ireland, 8 Castlecourt Centre, Dublin 15, D15 XTP3, Ireland (email: info@hbgi.ie)

Impression number 10 9 8 7 6 5 4 3

Year   2025

All rights reserved. Apart from any use permitted under UK copyright law, no part of this publication may be reproduced or transmitted in any form or by any means, electronic or mechanical, including photocopying and recording, or held within any information storage and retrieval system, without permission in writing from the publisher or under licence from the Copyright Licensing Agency Limited. Further details of such licences (for reprographic reproduction) may be obtained from the Copyright Licensing Agency Limited, www.cla.co.uk

Cover photo © Linden Gledhill

Illustrations by Aptara Inc

Typeset in Goudy Oldstyle Std 10/12 by Aptara Inc.

Printed and bound by CPI Group (UK) Ltd, Croydon, CR0 4YY

A catalogue record for this title is available from the British Library.

Contents

Introduction — iv

Studying IB Physics — viii

Experimental skills and abilities — 2

Chapter 1   Physical quantities and units — 4

Chapter 2   Practical techniques — 9

Chapter 3   Using apparatus — 26

Chapter 4   Mandatory practicals — 37

Chapter 5   Mathematical and measurement skills — 62

Chapter 6   Information communication technology (ICT) — 78

Writing the Internal Assessment report — 88

Chapter 7   Personal engagement — 90

Chapter 8   Exploration — 94

Chapter 9   Analysis — 107

Chapter 10   Evaluation — 115

Chapter 11   Communication — 123

Glossary — 130

Answers — 132

Index — 142

Introduction

How to use this book

There are two aspects to the practical work in the IB Diploma Programme Physics course: general practical work and a single individual investigation that forms the internal assessment (IA) project.

This publication is aimed specifically at IB Physics students and is to be used throughout your two years of study. Practical activities and the IA project form an essential part of the 2014 IB Diploma Programme Physics syllabus (first assessment held in 2016) with 40 hours recommended teaching time for standard level (SL) and 60 hours for higher level (HL). This represents an average 25% of the total teaching time. The internal assessment is worth 20% of the final assessment.

General practical work includes experiments in the laboratory, spreadsheets or online simulations (for example, Java Applets, Flash animations or Python (Trinket) simulations), demonstrations by your IB Physics teacher and class activities of a formative nature. These are designed to help you learn physics via practical work. Physicists use evidence gained from prior observations and experiments to build models and theories. Predictions are tested with practical work to check that they are consistent with the behaviour of the physical world.

The 'Applications and skills' section of the IB Diploma Programme Physics syllabus lists specific laboratory skills, techniques and experiments that you must experience at some point during your IB Physics course. Your school is likely to arrange additional practical work covering other topics in the IB Physics course. Note that it is the skills and not the specific experiments that will be assessed in the written examinations. Other recommended laboratory skills, techniques and experiments are listed in the 'Aims' section of the syllabus. Within the IB Physics syllabus, there is also a specific set of mandatory practicals that you should carry out over the course. Your knowledge and understanding of these will be assessed in your final examination papers.

This guide will ensure you can aim for your best grade by:

- building practical, mathematical and analytical skills for the mandatory and other common practicals through a comprehensive range of strategies and detailed examiner advice and expert tips

- offering concise, clear explanations of all the IB Diploma requirements, such as the assessment objectives of each assessment criterion for the IA, including checklists, and rules on academic honesty

- demonstrating what is required to obtain the best IA grade for the individual investigation with advice and tips, including common mistakes to avoid

- explicit reference to the IB learner profile and the associated approaches to learning (ATLs) that are central to the IB Diploma Programme, with their connections to practical work

- including exemplars with worked answers and commentary throughout, so you can see the application of physical and mathematical principles and concepts

- testing your comprehension of the skills covered with embedded activity questions (with answers at the back of the book).

Features of this book

> **Key definition**
>
> The definitions of essential key terms are provided on the page where they appear. These are words that you can be expected to know related to practical work and examination questions set in the context of experiments.

> **Examiner guidance**
>
> These tips give you advice that is likely to be in line with the thinking of IB Physics examiners.

> **Worked examples**
>
> Some practical skills require you to carry out mathematical calculations, plot graphs and so on. These examples show you how.

> ■ **ACTIVITY**
>
> Questions and suggested outlines of possible practice activities.

> **Investigations**
>
> Ideas for possible investigations.

> **RESOURCES**
>
> Useful websites or published books.

> **Expert tip**
>
> These tips give practical advice that will help you to boost your final IA grade.

> **Common mistake**
>
> These identify typical mistakes that IB Physics candidates may make and explain how you can avoid them.

Studying IB

IB Learner Profile

The IB Physics course is linked to the IB learner profile. Throughout the course, and while carrying out your internal assessment, you will have the opportunity to develop each aspect of the learner profile: Inquirers, Knowledgeable, Thinkers, Communicators, Principled, Open-minded, Caring, Risk-takers, Balanced and Reflective.

Practicals

Carrying out practicals throughout your IB Physics course will give you the opportunity to practice carrying out an investigation, and will give you the scientific skills you need for your internal assessment.

Studying IB Physics

Physics

Approaches to Learning

The IB Physics course, and the internal assessment in particular, give you the chance to develop the approaches to learning skills:
- thinking skills when planning investigations, collecting data and analyzing your results
- social skills when working with your peers
- communication skills when reporting and presenting your findings
- self-management skills when working independently
- research skills to help plan your investigation, and to put it into context.

Internal Assessment

The internal assessment gives you the opportunity to display the skills and knowledge you have learned throughout your course, while exploring an area of Physics that interests you personally.

Studying IB Phsyics

Physics and the scientific method

Physics is both an observational and an experimental science. Systematic observations and reliable measurements of physical phenomena may lead to a hypothesis. The hypothesis can lead to an experimental **investigation** that systematically manipulates a variable (under controlled conditions) in order to establish a relationship between two variables. This results in an improved understanding of physics. This **scientific method** can be seen as a cycle (Figure 1 and Table 1). Exploration of one idea can lead to further modification, through analysis and evaluation, resulting in the investigation and testing of further hypotheses.

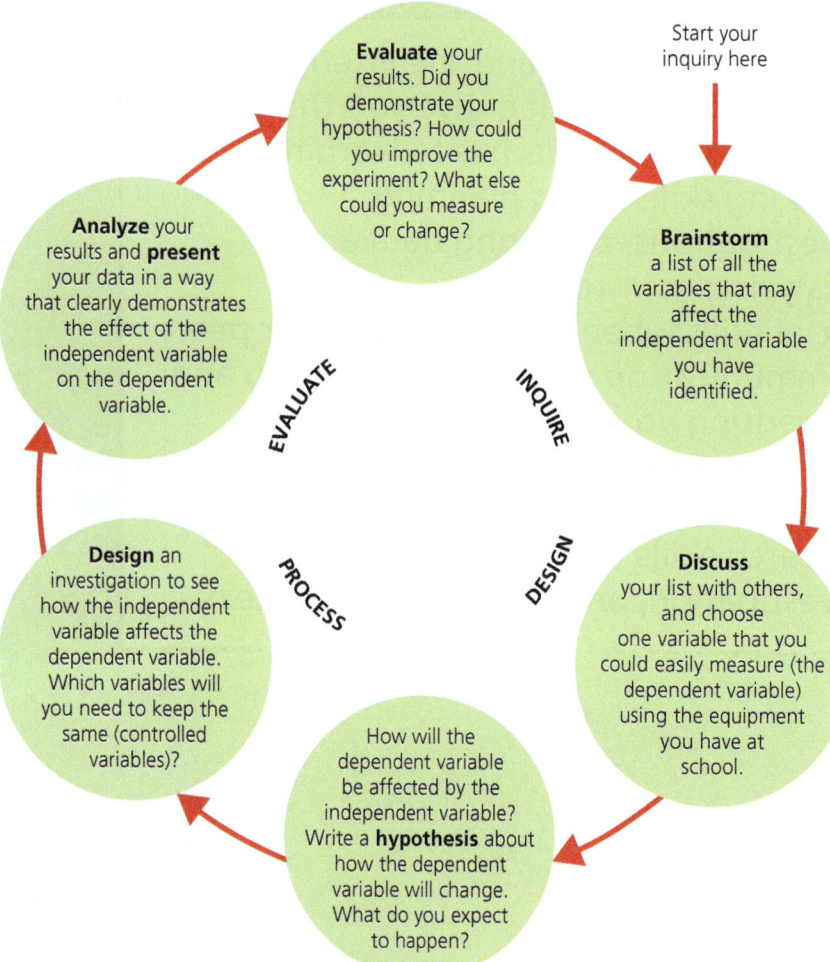

Figure 1 The investigation cycle

> **Key definitions**
>
> **Investigation** – A scientific study consisting of a controlled experiment in the laboratory.
>
> **Scientific method** – The use of controlled observations and measurements during an experiment to test a hypothesis.

Stage of cycle	Description	Key word definitions
1	A research question is formulated which usually enquires how one variable (the independent variable) affects another (the dependent variable). For example, during an investigation a thermistor may be heated to different temperatures (independent variable) and its resistance measured (dependent variable).	**Variable** – A condition or factor that can vary and may be varied during an investigation and is likely to affect the value of a related quantity. **Independent variable** – A variable whose value is changed (across a range) by the experimenter in an investigation, to establish its effect on the dependent variable. **Dependent variable** – The variable that is being directly measured in an investigation. Its value depends on the independent variable.
2	The research question may lead to the formulation of a hypothesis: the investigator proposes and explains how (via a scientific model) the independent variable may affect the dependent variable (in a causal relationship). A prediction, ideally quantitative, may be made.	**Hypothesis** – A proposed explanation based on limited data and observations, the predictions of which may be tested by investigation. **Prediction** – Expected results of an investigation.
3	So that a correlation, ideally a quantitative relationship, between the two variables (independent and dependent) can be established, other variables must be kept the same and monitored – these are called controlled variables. They should be listed, and information about why and how they will be kept constant included.	**Quantitative relationship** – A relationship between variables that can be described by a mathematical equation. **Controlled variables** – Variables not under investigation that are kept constant and monitored during an experiment.
4	A method is developed and clearly outlined. A clearly labelled cross-sectional diagram should be included showing the relative positions of the apparatus and instruments. One variable should be manipulated and one measured, with other variables controlled (and monitored). The adopted method should be written clearly in enough detail so that someone else can follow the instructions and obtain reproducible results.	
5	The investigation is carried out and raw data (typically instrument readings) are collected by measurement of the dependent variable. Data should be recorded in tables with appropriate units (usually SI base units and SI derived units) and uncertainty ranges. Raw data will often be processed, for example, means, squares, reciprocals or logarithms calculated. The new values form processed data. A graph (typically a line graph with a line of best fit) should be plotted from ordered pairs of measured values of the independent and dependent variables. The graph should present the trend clearly and help mathematical analysis (for example, by determining the gradient or intercept). Any non-linear relationships should be linearized (where possible and appropriate). Relevant qualitative data (observations) may also be recorded.	**Raw data** – Recorded observations and measurements that have not yet been processed or analysed. **Measurement** – A record of, or the process of recording, the size of a physical quantity: its numerical amount and its unit. **Processed data** – Raw experimental data that has been mathematically manipulated. **Linearization** – Converting a relationship to a linear one by mathematically modifying the variables.
6	Following analysis, an explanation for the results is developed. A written description details what the raw, processed and displayed data show about the relationship between the variables. The results (data) are assessed with reference to the research question and the relevant scientific model. It should be stated whether the results support the hypothesis, or falsify it.	**Analysis** – Recognizing and commenting on trends in raw and processed data and stating valid conclusions.
7	An evaluation is made of the investigation. Its limitations are commented on, considering the procedures, the instruments, their sensitivity, their use, the quality of the data (their accuracy and precision), the overall uncertainty in the data and the reliability of the results. The extent to which the limitations might have affected the results is considered. Realistic improvements or extensions to the investigation are proposed that address the limitations and increase accuracy or precision.	**Evaluation** – An assessment of all the limitations of an investigation, the quality of the data and the reliability of the conclusions. **Sensitivity** – The smallest change that can be detected by a measuring instrument. **Accuracy** – How close a measured value is to the true or theoretical value. **Precision** – How close repeated measurements of the same quantity are to their mean value.
8	The suggested improvements and extensions to the method can lead to further investigations, and so the cycle (of the scientific method) repeats itself.	

Table 1 The scientific method cycle

The nature of physics

In physics, many of the concepts studied, such as energy and fields, are not directly observable. Physicists often make use of a simplified model of a system to help understand and predict physical phenomena. Physics uses mathematics to describe such systems. The results of practical investigations are essential in establishing such mathematical descriptions. These quantitative data (measurements) show the level of agreement between the predictions about the actual behaviour of a system in the physical world.

Replication of experimental measurements can improve the **reliability** of the data generated by an investigation and enable **anomalous data** to be identified.

Data are reliable if consistent values are obtained when the measurement is repeated. Data should also be valid, which means the measurements are of the required data or can be processed to give the required data to resolve the research question.

Data can be either qualitative or quantitative. **Qualitative data** are recorded observations not involving measurements, for example, changes in the observed viscosity of a liquid or the heating of a resistor during an experiment. **Quantitative data** are numerical data, with units and ideally with their **uncertainty** stated, from measurements taken when recording a dependent variable (such as the value of the current through a light dependent resistor).

Raw data collected may be difficult to use for data analysis, and often need to be processed in some way by carrying out a calculation. A processed variable is one that is calculated from raw data (often from measurement of the dependent variable). For example, in a practical to determine the force (spring) constant of a spring the dependent variable may be the time for 20 oscillations and the processed variable may be the (mean) time for one oscillation which is then squared (in an attempt to verify a linear relationship with the independent variable, mass).

A **continuous variable** can take any numerical value, for example, the release height of a ball bearing can vary continuously. The data can be represented as a line graph. A **discrete variable** has only a certain number of allowed values, or types or categories, such as type of material. Discrete data can only be presented as a column graph, not as a line graph.

> **Examiner guidance**
>
> A strong **correlation** between two variables does not necessarily mean causation. For example, a stretching force causes *both* a tension *and* an extension and these two variables are correlated but not causally related.
>
> In a general case where X and Y are an independent and a dependent variable, you cannot state definitively that a change in X causes the changes in Y. In a particular case you may *suggest* that a change in X causes changes in Y, or vice versa. However, you should not draw definite 'cause and effect' conclusions based on correlation. There are several reasons for this:
>
> - You may not know the direction of the cause – does X cause Y, or does Y cause X?
>
> - A third variable Z may be involved that is responsible for the correlation between X and Y. For example, increasing mass could overstretch the spring and affect the force constant (otherwise assumed to be constant).
>
> - The apparent relationship may simply be due to chance (although this is highly unlikely in a well planned investigation).

> **Key definition**
>
> **Replication** – A repeating of the entire experiment (on the same occasion with the same apparatus) to record repeat measurements and observations.
>
> **Reliability** – The extent to which the results of an investigation can be consistently replicated, within limits of experimental uncertainty.
>
> **Anomalous data** – Data with unexpected values (beyond the limits expected by uncertainty) that do not match the relationship predicted by the hypothesis, or the trend shown by the rest of the data.
>
> **Qualitative data** – Observations recorded without measurements being taken.
>
> **Quantitative data** – Numerical results; measurements with units.
>
> **Uncertainty** – Defined range of measured values in which the true value is the central point.
>
> **Continuous variable** – A variable that can take any numerical value within a range.
>
> **Discrete variable** – A variable that can only have a certain number of values (that do not have an order), or categories.
>
> **Correlation** – When one variable appears to have a relationship with another.

Framework for physics

The IB Physics syllabus is comprehensive and detailed, but it may be helpful to simplify the content by using a 'concept map' (Figure 2) that outlines the essential components and concepts of your course and how they interrelate. The practical skills you will be learning during the course can be framed in the context of this diagram. Subsequent chapters will cover essential skills that you can apply in different areas of the syllabus. You can also use Figure 2 to help you choose the area of physics that you want to address in your internal assessment (IA) project (Chapter 8).

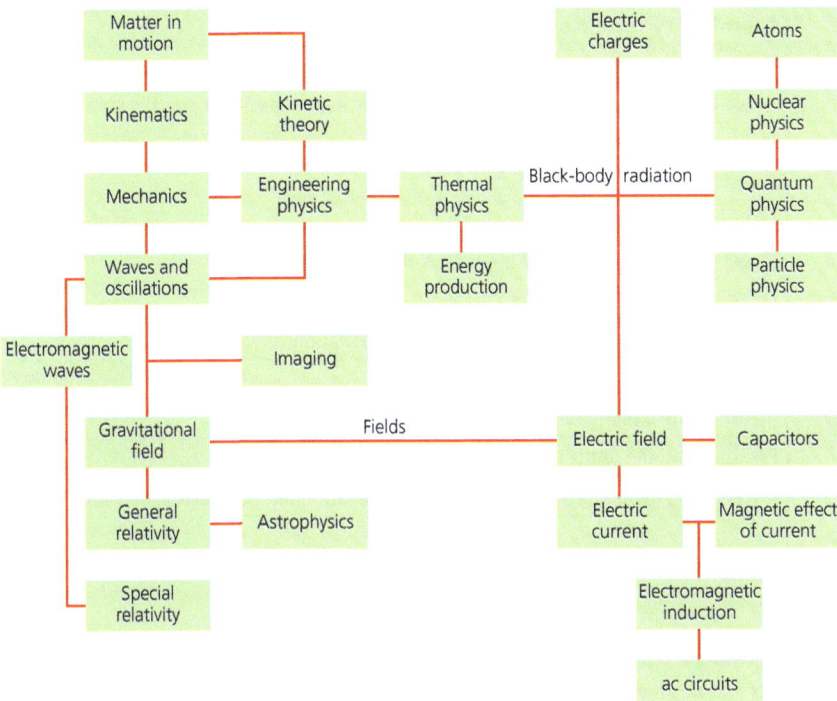

Figure 2 Physics concept tree

IB Physics practicals

There are numerous practicals listed in the *IB Physics Guide*. This book focuses on the practical experiments that can help you not only with exam questions (Paper 3, section A) but also in the selection and implementation of a suitable investigation for your internal assessment (IA). The practicals in the *IB Physics Guide* can be divided into three categories.

- Mandatory practicals: these are prescribed and you need an understanding of these experiments because they may be examined in Paper 3.

- Useful additional suggested practicals, to be chosen at the discretion of your physics teacher. These will not be specifically examined, but may provide useful ideas to help you select and then implement your IA project.

- Computer simulations: information communication technology (ICT) is encouraged throughout all aspects of your physics course. Certain ICT skills are specified in the *IB Physics Guide*, which involve the use of computers to model or draw associations between numerical data.

> **Examiner guidance**
>
> **Secondary data** are data obtained from another source, such as via reference material or published third-party results. Analysis of secondary data can form an important part of experimental work, for example, spectral data (https://avogadro.cc/docs/tutorials/viewing-vibrational-spectra/) and images of receding galaxies are available from professional astronomical observatories for analysis. The sources of secondary data must always be cited in a report.

> **Key definitions**
>
> **Secondary data** – data obtained from another source, such as via reference material or published third-party results. Analysis of secondary data can form an important part of experimental work.

Mandatory practicals

Subtopic	Mandatory practical
2.1 Motion	Determining the acceleration of free-fall
3.1 Thermal concepts	Applying the calorimetric techniques of specific heat capacity or specific latent heat
3.2 Modelling a gas	Investigating at least one gas law
4.2 Travelling waves	Investigating the speed of sound
4.4 Wave behaviour	Determining refractive index
5.2 Heating effect of electric currents	Investigating one or more of the factors that affect resistance
5.3 Electric cells	Determining internal resistance
7.1 Discrete energy and radioactivity	Investigating half-life experimentally (or by simulation)
9.3 Interference [HL only]	Investigating Young's double slit
11.2 Power generation and transmission [HL only]	Investigating a diode bridge rectification circuit
C.2 Imaging instrumentation [Option C only]	Investigating the optical compound microscope
C.2 Imaging instrumentation [Option C only]	Investigating the performance of a simple optical astronomical refracting telescope

Table 2 Mandatory practicals for physics

Suggested practicals

The following suggested practicals will help you to enhance your understanding of physics or could be individual investigations. Your physics teacher may select them to include in your practical scheme of work (PSOW). They are listed in the 'Guidance' section of the syllabus (under 'Aims') and will not be examined.

Subtopic	Suggested practical
2.1 Motion	Estimating speed using travel time tables
	Investigating motion through a liquid (http://practicalphysics.org/falling-through-water.html)
	Analysing projectile motion (https://tap.iop.org/mechanics/kinematics/207/file_46334.doc)
2.2 Forces	Verification of Newton's second law (http://practicalphysics.org/force-mass-and-acceleration-newtons-second-law.html)
	Investigating forces in equilibrium
	Determination of the effects of friction
2.3 Work, energy and power	Relationship of kinetic and gravitational potential energy for a falling mass (http://www.schoolphysics.co.uk/age16-19/Mechanics/Rotation%20of%20 rigid%20bodies/experiments/moment_of_inertia_of_a_flywheel.doc)
	Power and efficiency of mechanical objects
	Comparison of different situations involving elastic potential energy
2.4 Momentum and impulse	Analysis of collisions with respect to energy transfer
	Impulse investigations to determine velocity, force, time, or mass
	Determination of amount of transformed energy in inelastic collisions (http://practicalphysics.org/momentum.html)
3.1 Thermal concepts	Transfer of energy due to temperature difference
	Calorimetric investigations (https://tap.iop.org/energy/thermal/607/page_47500.html)
	Energy involved in phase change (https://tap.iop.org/energy/thermal/608/page_47512.html)
3.2 Modelling a gas	Verification of a gas law
	Calculation of the Avogadro constant (https://www.lahc.edu/classes/chemistry/arias/Exp%203%20-%20 AvogadroF11.pdf)
4.1 Oscillations	Mass on a spring (http://practicalphysics.org/investigating-mass-spring-oscillator.html)
	Simple pendulum (https://tap.iop.org/vibration/shm/304/page_46587.html)
	Motion on a curved air track (http://users.df.uba.ar/sgil/physics_paper_doc/papers_phys/mechan/airdrag1.pdf)

Topic	Suggested practicals
4.2 Travelling waves	Speed of waves in different media
	Detection of electromagnetic waves from various sources
	Use of echo methods (or similar) for determining wave speed, wavelength, distance, or medium elasticity and/or density
4.3 Wave characteristics	Observation of polarization under different conditions, including the use of microwaves
	Superposition of waves (http://tap.iop.org/vibration/superpostion/index.html)
	Representation of wave types using physical models (for example, slinky demonstrations)
4.5 Standing waves	Observation of standing wave patterns in physical objects (for example, slinky springs)
	Prediction of harmonic locations in an air tube in water
	Determining the frequency of tuning forks
	Observing or measuring vibrating violin/guitar strings (http://practicalphysics.org/measuring-speed-sound-using-echoes.html)
5.1 Electric fields	Demonstrations showing the effect of an electric field (for example, using semolina) (http://practicalphysics.org/electric-fields.html) (https://www.andrews.edu/phys/wiki/PhysLab/doku.php?id=lab1)
5.2 Heating effect of electric currents	Use of a hot-wire ammeter as an historically important device (http://practicalphysics.org/model-hot-wire-ammeter.html)
	Comparison of resistivity of a variety of conductors such as a wire at constant temperature, a filament lamp, or a graphite pencil
	Determination of thickness of a pencil mark on paper (http://www.kfupm.edu.sa/sites/phys102/PDF/thinSample.pdf)
	Investigation of ohmic and non-ohmic conductor characteristics
	Using a resistive wire wound and taped around the reservoir of a thermometer to relate wire resistance to current in the wire and temperature of wire
5.3 Electric cells	Investigation of simple electrolytic cells using various materials for the cathode, anode and electrolyte (http://practicalphysics.org/electrolysis.html)
	Comparison of the life expectancy of various batteries
6.1 Circular motion	Investigation of mass on a string
	Observation and quantification of loop-the-loop experiences
	Friction of a mass on a turntable (http://universe.bits-pilani.ac.in/uploads/Goa-Physics/_pdf_archive_AJPIAS_vol_83_iss_2_126_1.pdf)
9.1 Simple harmonic motion	Investigation of simple or torsional pendulums (http://www.schoolphysics.co.uk/age16-19/Mechanics/Rotation%20of%20rigid%20bodies/text/Rotation_of_rigid_bodies/index.html)
	Measuring the vibrations of a tuning fork (http://practicalphysics.org/examples-simple-harmonic-motion.html)
9.3 Interference	Observing the use of diffraction gratings in spectroscopes
	Analysis of thin soap films (http://practicalphysics.org/soap-film.html)
	Sound wave and microwave interference pattern analysis (http://practicalphysics.org/interference-using-centimetre-waves.html)
11.2 Power generation and transmission	Construction of a basic ac generator
	Investigation of variation of input and output coils on a transformer (http://practicalphysics.org/oscilloscope-and-alternating-voltage-transformer.html)
	Observing Wheatstone and Wien bridge circuits (http://physics.ucdavis.edu/Classes/Physics116/Lab10_rev2.pdf)
11.3 Capacitance	Investigating basic RC circuits (https://tap.iop.org/electricity/capacitors/126/page_46162.html)
	Using a capacitor in a bridge circuit
	Examining other types of capacitors
	Verifying the time constant for a capacitor (https://tap.iop.org/electricity/capacitors/129/page_46197.html)
12.1 The interaction of matter with radiation	Investigation of the photoelectric effect using LEDs (http://media4.physics.indiana.edu/~courses/p309/exp-procedure/Photoelectric%20Effect/garver-PhotoElectricEffect.pdf)
B.4 Forced vibrations and resonance	Observation of sand on a vibrating surface of varying frequencies (http://www.physicsclassroom.com/class/sound/Lesson-4/Standing-Wave-Patterns)
	Investigation of the effect of increasing damping on an oscillating system, such as a tuning fork
	Observing the use of a driving frequency on forced oscillations (https://tap.iop.org/vibration/shm/307/page_46612.html)
C.1 Introduction to imaging [Option C only]	Magnification determination using an optical bench
	Investigating real and virtual images formed by lenses (http://www.nuffieldfoundation.org/topic/141/933)
	Observing aberrations (http://www.physics.louisville.edu/sbmendes/phys%20356%20fall%202013/Aberration%20of%20a%20Lens.pdf)

Table 3 Suggested practicals

Computer simulations

Other practical skills involve the use of ICT (Table 4).

Subtopic	Activity/simulation
2.1	Simulations of terminal velocity (http://www.physicsclassroom.com/Teacher-Toolkits/Terminal-Velocity/Terminal-Velocity-Complete-ToolKit)
2.4	Simulations of molecular collisions (http://www.falstad.com/gas/)
3.2	Simulation of a gas law (https://phet.colorado.edu/en/simulation/legacy/gas-properties) (https://ch301.cm.utexas.edu/section2.php?target=gases/kmt/gas-simulator.html)
4.3	Simulating waves in three dimensions (http://www.falstad.com/wavebox/) and wave superposition (http://iwant2study.org/lookangejss/04waves_11superposition/ejss_model_wave1d01/wave1d01_Simulation.xhtml)
5.1	Simulations involving the placement of one or more point charges and determining the resultant field (https://phet.colorado.edu/en/simulation/charges-and-fields)
5.2	Simulation of electric circuits (http://www.physicsclassroom.com/Physics-Interactives/Electric-Circuits/Circuit-Builder)
5.4	Simulations of electromagnetic fields in three-dimensional space (https://phet.colorado.edu/en/simulation/legacy/radio-waves)
7.1	Investigating half-life experimentally (or by simulation) (http://www.nuffieldfoundation.org/practical-physics/simple-model-exponential-decay)
7.2	Investigating the scattering angle of alpha particles as a function of the aiming error, or the minimum distance of approach as a function of the initial kinetic energy of the alpha particles (https://tap.iop.org/atoms/rutherford/index.html)
8.2	Simulations of energy exchange in the Earth surface–atmosphere system (https://phet.colorado.edu/en/simulation/legacy/greenhouse)
9.1	Simulation of simple harmonic motion (https://www.geogebra.org/m/pY4Hvugh)
9.5	Simulation of the Doppler effect (https://highered.mheducation.com/olcweb/cgi/pluginpop.cgi?it=swf::800::600::/sites/dl/free/0072482621/78778/Doppler_Nav.swf::Doppler%20Shift%20Interactive)
B1	Simulation of outcome of actions on bodies (https://phet.colorado.edu/en/simulation/legacy/torque) (https://phet.colorado.edu/en/simulation/rotation)
B3	Simulation of fluid dynamics phenomena (https://physics.weber.edu/schroeder/fluids/)
B4	Simulation of resonance in electrical circuits, atoms/molecules or radio/telecommunications using software (https://phet.colorado.edu/en/simulation/legacy/resonance)

Table 4 Computer activities

Approaches to learning

Approaches to learning (ATLs) are deliberate strategies, skills and attitudes which underlie all aspects of the IB Diploma Programme. These approaches are intrinsically linked with the IB Learner profile attributes (see below) and are designed to enhance your learning and preparation for the IB Diploma Programme assessment and beyond.

> **Expert tip**
>
> ATLs encompass the key values and principles that underpin an IB education.

The aims of ATLs in the Diploma Programme are to:

- link prior knowledge to course-specific understandings, and make connections between different subjects
- encourage you to develop a variety of skills that will equip you to continue to be actively engaged in learning after you leave your school or college
- help you not only to obtain university admission through better grades but also to prepare for success during tertiary education and beyond
- enhance further the coherence and relevance of your IB Diploma Programme experience.

The five approaches to learning develop the following skills:

- thinking skills
- social skills
- communication skills
- self-management skills
- research skills.

Practical activities clearly allow you to interact directly with and scientifically study natural phenomena, explore a topic and examine specific research questions. All practical skills covered in this book can be viewed in the context of ATLs. They also give you the opportunity to develop and use IB terminology:

- research skills to find out appropriate methods to investigate specific research questions, and put your investigation in the context of the wider scientific community
- thinking skills to design investigations, collect and analyse data, and then evaluate your results
- social skills in order to collaborate with peers
- communication skills to effectively and concisely present your findings
- self-management skills to make sure you successfully plan your time.

The IB learner profile

The IB Diploma Programme physics course is closely linked to the IB learner profile. By following the course, you will have engaged with all attributes of the IB learner profile, and the requirements of the internal assessment provide opportunities for you to develop every aspect of the profile (Table 5).

IB learner profile attribute	Relevance to IB Physics syllabus
Inquirers	Practical work and internal assessment
Knowledgable	Links to international-mindedness
	Practical work and internal assessment
Thinkers	Links to theory of knowledge (TOK)
	Practical work and internal assessment
Communicators	External assessment (examinations)
	Practical work and internal assessment
Principled	Practical work and internal assessment
	Ethical behaviour and academic honesty
Open-minded	Links to international-mindedness
	Practical work and internal assessment
	The group 4 project
Caring	Practical work and internal assessment
	The group 4 project
	Ethical behaviour
Risk-takers	Practical work and internal assessment
	The group 4 project
Balanced	Practical work and internal assessment
	The group 4 project
	Field work
Reflective	Practical work and internal assessment
	The group 4 project

Table 5 Relevance of the IB learner profile to the IB Physics syllabus

The internal assessment

The internal assessment (IA) forms 20% of your final mark, with the external examinations (Papers 1, 2 and 3) forming 80% of your mark. The assessment and the assessment criteria are the same for both standard level (SL) and higher level (HL) physics.

Criterion	Personal engagement	Exploration	Analysis	Evaluation	Communication	Total marks available
Marks available	2	6	6	6	4	24

Table 6 Marking criteria for the physics IA

Your IA mark is based on one scientific investigation, known as the individual investigation. This will involve 10 hours of work and generating a word-processed report or write-up between 6 to 12 pages. This will be marked out of a maximum of 24 marks based upon the five group 4 assessment criteria (Table 6). This will then be scaled to a mark out of 20. Your individual investigation will be internally marked by your IB Physics teacher but moderated externally (re-marked) by an experienced IB Physics teacher appointed by the IBO.

There are separate chapters in this book for each of the IA criteria (Chapters 7–11). Checklists at the end of each criterion chapter will be helpful in ensuring that your report matches the requirements of the group 4 assessment criteria.

Grade boundaries for the IA are as follows (using data from examinations from May 2016 to May 2017).

Grade	1	2	3	4	5	6	7
Mark range	0–3	4–6	7–10	11–13	14–16	17–19	20–24

Table 7 Grade boundaries for the physics IA

■ Planning an IA

There are no IB requirements in terms of planning, a timeline or documentation, but your school may require you to complete a preliminary IA proposal in which you suggest a research question and **methodology** and carry out a **risk assessment**. You may also be asked to complete a requisition for apparatus, instruments and materials for preliminary work.

> **Key definitions**
>
> **Methodology** – The methods/techniques used to carry out an investigation and their justification.
>
> **Risk assessment** – A consideration of the hazards that impact human health that could be encountered during an investigation and their level of risk, as well as any environmental impact (for example of disposal).

■ Setting up a schedule

You may find it helpful to set up a timeline with start dates and deadlines for each part of your individual investigation, if your school has not done that. A sample timeline is shown in Table 8.

	Start date	Task	Deadline date
Planning 1		Read Chapters 7 and 8 in this book.	
Planning 2		Decide on the research question, identify and classify all the variables, formulate a hypothesis (if appropriate), outline your methodology and data collection and processing.	
Planning 3		Prepare a risk assessment for these experiments and show your physics teacher the completed risk assessment form.	
Planning 4		Check that the apparatus, instruments and materials will be available in your school physics laboratory.	
Practical		Complete the experimental work safely and collect the raw data in the time allocated. Allow sufficient time to carry out replicates, extending the range of data collected, and for any preliminary work. Document any alterations to your plan as soon as they occur, and if necessary make alterations to the supporting theory.	
Report 1		Hand in the first draft and consult with your physics teacher.	
Report 2		Submit the final draft after an online plagiarism check.	

Table 8 Suggested timeline for individual investigation

> **Expert tip**
>
> You may want to use a bound and numbered exercise book as a scientific log book to record all that happened during your individual investigation. Use it for thinking, calculating, drawing, recording raw data (including observations) and planning your IA report. Your log book may include initial ideas, notes from background reading, data analysis and any difficulties you encountered.

Experimental skills

- Physical quantities and units
- Scientific notation
- Measurement
- Errors
- Graphs

Mathematical and measurement skills

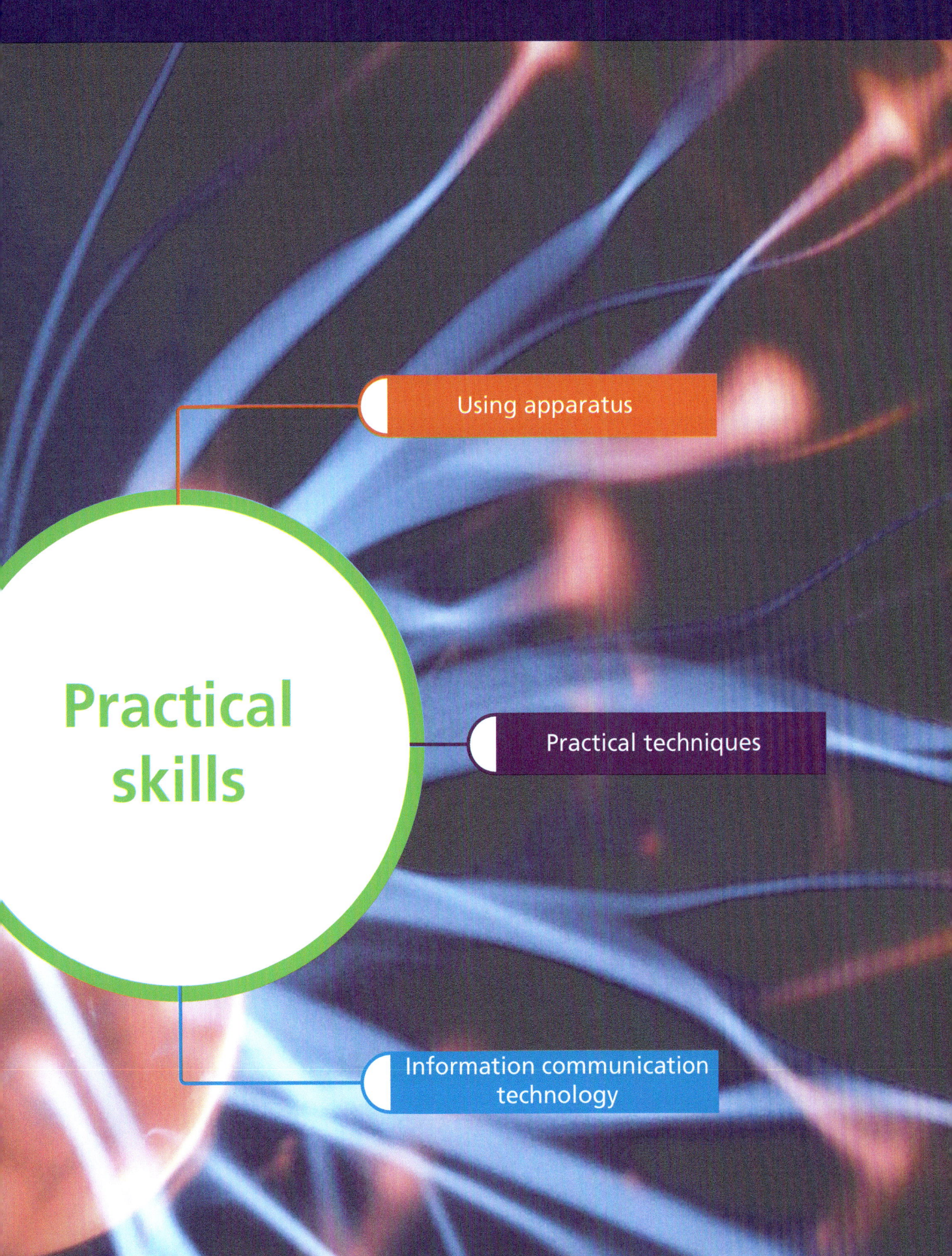

1 Physical quantities and units

Units

A physical property that can be quantified by measuring is a **physical quantity**. For example, the length of a rod and the mass of a trolley are physical quantities. When recording the measurement of a physical quantity you should include the numerical value of the quantity (its magnitude), its unit of measurement and its uncertainty.

> **Key definition**
>
> **Physical quantity** – Property of an object that can be quantified (given a numerical value).

> **Worked example**
>
> A force measurement might be 5.0 N ± 0.5 N. Here, 5.0 is the numerical value of the quantity, the newton (N) is the unit of measurement, and the uncertainty in the value is plus or minus 0.5 N. This implies that the true value of the quantity lies between 5.5 N and 4.5 N.

Fundamental units

The SI system of units (Table 1.1) is based on seven independent fundamental (base) units used with the decimal number system.

Fundamental or base physical quantity		SI base unit	
Name	Symbol	Name	Symbol
Length	l or x	metre	m
Mass	m	kilogram	kg
Time	t	second	s
Electric current	I	ampere	A
Temperature	T	kelvin	K
Amount of substance	n	mole	mol
Luminous intensity	I_v	candela	cd

Table 1.1 Fundamental or base SI units

> **Examiner guidance**
>
> The symbol for a physical quantity is a single letter in italics, for example, t indicates time. The symbol for a unit is not in italics. Unit symbols derived from a name are a capital letter, or begin with a capital letter (although the unit is not written with a capital letter, for example, J for joule). All other unit symbols use lower case letters, for example, m for metre. The symbols are unchanged in singular or plural and do not have a full stop after them.

Pure numbers and numerical constants (for example, π) have no unit; they are dimensionless. Some physical quantities, for example, refractive index, are also dimensionless quantities, because they are ratios; these have no unit. However, not all physical constants are dimensionless, for example, the speed of light, c, has unit m s^{-1} and the acceleration due to free fall, g (on or near the Earth's surface), has unit m s^{-2}.

> **Examiner guidance**
>
> Do not make up your own symbols for physical quantities, for example, c to represent current. Important discoveries about electricity were made by the French physicist Ampère; he wrote about the 'intensity' of electric current (*intensité de courant*) and to honour his work the symbol I represents current.

Derived units

Mensurationally derived quantities are defined in terms of the SI base quantities via a quantity equation. (Mensuration refers to the act of measuring.) The SI derived units for these derived quantities can be obtained from these equations and the seven SI base units (Table 1.2).

Mensurationally derived physical quantity	SI derived unit	
	Name	Symbol in terms of SI base units
Volume	cubic metre	m^3
Density	kilogram per cubic metre	$kg\ m^{-3}$
Velocity	metre per second	$m\ s^{-1}$
Acceleration	metre per second squared	$m\ s^{-2}$
Specific heat capacity	joule per kilogram per kelvin	$J\ kg^{-1}\ K^{-1}$

Table 1.2 Selected derived SI units

Examiner guidance

The term 'specific' refers to 'divided by mass', for example, specific heat capacity is the heat capacity per kilogram of the material. The SI unit of specific heat capacity is joule per kilogram per kelvin ($J\ kg^{-1}\ K^{-1}$).

■ ACTIVITY
1. A piece of nickel wire is 2.0 cm long with a diameter of 0.25 mm. Calculate the volume of the nickel wire in m^3.

The base and derived SI units form a coherent system of units. All the units for the derived physical quantities are obtained from the base units by multiplication or division without introducing numerical factors, which simplifies many calculations. For example, consider the ideal gas equation, $PV = nRT$. If numerical values of V, n, R and T are substituted into the equation using SI units, the calculated value of P will be in pascals, the SI derived unit for pressure, equivalent to $N\ m^{-2}$.

Some units of other derived quantities have been given special names (Table 1.3).

Mensurationally derived physical quantity	SI derived unit		Equivalent in terms of SI base units
	Name	Symbol	
Force	newton	N	$m\ kg\ s^{-2}$
Energy and work	joule	J	$m^2\ kg\ s^{-2}$
Power	watt	W	$m^2\ kg\ s^{-3}$
Frequency	hertz	Hz	s^{-1}
Pressure	pascal	Pa	$kg\ m^{-1}\ s^{-2}$
Radioactivity	becquerel	Bq	s^{-1}
Quantity of charge	coulomb	C	$A\ s$
Electric potential difference, electromotive force	volt	V	$m^2\ kg\ s^{-3}\ A^{-1}$
Electrical resistance	ohm	Ω	$m^2\ kg\ s^{-3}\ A^{-2}$
Magnetic field strength	tesla	T	$kg\ s^{-2}\ A^{-1}$
Magnetic flux	weber	Wb	$kg\ m^2\ s^{-2}\ A^{-1}$
Dose	gray	Gy	$m^2\ s^{-2}$

Table 1.3 Selected derived units with names

You must recognise and use derived units in the negative exponent form, for example, $m\ s^{-1}$ rather than m/s.

Use of dimensions to derive equations

Dimensions in physics are the powers to which the fundamental quantities of mass, length and time must be raised to represent a physical quantity. The dimension of mass is denoted as [M], that of length [L] and that of time [T]. The dimensions of speed will then be $[L][T]^{-1}$ and those of density will be $[M][L]^{-3}$.

A valid equation in physics must be dimensionally homogenous, that is it must have the same dimensions on both sides, since equality cannot apply between quantities of different nature.

Dimensional analysis can be used to obtain an equation relating relevant variables in a practical investigation. Consider the oscillation of a simple pendulum. It can be assumed that the period, T, will vary with the mass of the pendulum, m, length of the string, l, and gravitational field strength, g.

The equation can therefore be written in the following form:

$T = k\, m^x\, l^y\, g^z$

where x, y and z are unknown powers and k is a dimensionless constant. Writing this equation in dimensional form gives:

$[T] = [M]^x [L]^y [L]^z [T]^z$

This can be transformed to:

$[T] = [M]^x [L]^y [L]^z [T]^{-2z}$

since g has unit m s^{-2}.

Equating the indices for M, L and T on both sides of the dimensional equation above (to ensure homogeneity) gives:

M: $0 = x$; L: $0 = y + z$; T: $1 = -2z$

and hence $x = 0$; $y = \frac{1}{2}$ and $z = -\frac{1}{2}$. The original equation for the period, T, therefore becomes:

$T = k\sqrt{\left(\dfrac{l}{g}\right)}$

The constant, k, can be shown, by another method, to be 2π. (If the amplitude of the motion of the swinging pendulum is small, then the pendulum behaves approximately as a simple harmonic oscillator. Simple harmonic motion is a motion that experiences a restoring force proportional to the displacement of the system.)

Ideas for investigations

There are a variety of pendulum types that are suitable systems for investigating and measuring g: simple, compound, Katers' and bifilar. A ball bearing undergoing small oscillations on a concave mirror can also be used to determine g.

Examiner guidance

Dimensional analysis can indicate if an equation is wrong, but it does not necessarily lead you to the correct equation. Equations that are homogeneous may or may not be correct. Possible causes of equations that are homogeneous but incorrect include: the presence/absence of a dimensionless constant (for example, π), an incorrect coefficient, or the presence of extra (but less significant) terms.

■ ACTIVITY

2 Use dimensional analysis to suggest an equation relating the speed of waves, v, on a string with the mass of the string m, length, l, and tension force, T.

Metric multipliers

One advantage of the SI system is its use of standard prefixes or metric multipliers (Table 1.4) to represent multiples of 10, so that readings of very large or very small values can be easily expressed.

Prefix	Multiplying factor			Symbol
	in figures	in scientific notation	in words	
giga	1 000 000 000	10^9	1 billion	G
mega	1 000 000	10^6	1 million	M
kilo	1000	10^3	1 thousand	k
deci	0.1	10^{-1}	1 tenth	d
centi	0.01	10^{-2}	1 hundredth	c
milli	0.001	10^{-3}	1 thousandth	m
micro	0.000 001	10^{-6}	1 millionth	µ
nano	0.000 000 001	10^{-9}	1 billionth	n

Table 1.4 Standard unit prefixes

The prefixes increase or decrease by a factor of a thousand, so by choosing a suitable prefix, all values can be in the range 1 to 999, for example, 2.57 MPa rather than 2 570 000 Pa. Use a space to indicate groups of thousands, not a comma, for example, 73 000 not 73,000.

> **Examiner guidance**
>
> The prefixes are written with no space between the letters, for example, kN for kilonewtons, or nm for nanometres. Notice that some prefixes are capital letters and some are small letters. K is the unit for temperature and M is the prefix for mega. Do not use Kg to represent the kilogram (kg) nor M for metre (m).

Converting units

Converting between unit prefixes involves multiplying appropriately. When converting a quantity from a factor of 10^a to 10^b, the quantity is multiplied by 10^{a-b}. For example, converting 8 mm to µm requires a multiplication by $10^{-3--6} = 10^3$, hence 8 mm = 8×10^3 µm.

ACTIVITIES

3   Rewrite the following quantities using suitable prefixes:
    0.000 000 650 m   3000 m   82 000 g   123 000 N
    950 000 000 Hz   0.08 A

4   Deduce which is larger, 0.167 GW or 1500 MW.

> **Worked example**
>
> Convert 500 m s^{-1} into km h^{-1}.
>
> Rewriting the physical quantity, multiplying by the unit conversion factors:
>
> $$500 \text{ m s}^{-1} = \frac{500 \times \frac{1}{1000} \text{ km}}{\frac{1}{3600} \text{ h}} = 1800 \text{ km h}^{-1}$$

> **Common mistake**
>
> Errors often occur when converting areas and volumes. Consider a square with area 10 cm × 10 cm = 100 cm². A common mistake is to calculate the area incorrectly as 100×10^{-2} m². You should convert to base units first:
>
> 10×10^{-2} m × 10×10^{-2} m = 100×10^{-4} m²
>
> Similarly, for volume there are 1×10^6 cm³ in 1 m³ because
> 1 m³ = 10^2 cm × 10^2 cm × 10^2 cm = 10^6 cm³.

ACTIVITY

5 Convert:

   a 2.20 m³ to cm³

   b 5900 mm² to cm²

   c 0.046 km³ to m³

   d 90 km h⁻¹ into m s⁻¹.

   Give your answers in scientific notation (see Chapter 5) where appropriate.

Non-SI units

Table 1.5 shows a selection of non-SI units that you may encounter during your IB Diploma Programme physics course.

Quantity	Alternative unit	Unit symbol	Value in SI units
Energy	electronvolt	eV	1.60×10^{-19} J
Temperature	degree Celsius	°C	$t/°C = T/K - 273.15$
Energy	kilowatt-hour	kWh	3.60×10^{6} J
Charge	elementary charge	e	1.60×10^{-19} C
Pressure	atmosphere	atm	$\approx 10^{5}$ Pa
Mass	unified atomic mass unit	u	1.661×10^{-27} kg
Time	hour	h	3600 s
Time	year	yr	31.6×10^{7} s
Distance	light year	ly	9.46×10^{15} m
Distance	parsec	pc	3.10×10^{16} m
Distance	astronomical unit	AU	1.50×10^{11} m

Table 1.5 Non-SI units

> **Expert tip**
>
> When you are converting from one of these non-SI units to SI units you need to multiply by the value in the right-hand column. When you convert back you need to divide.

ACTIVITIES

6 The nearest star (other than the Sun) to Earth is *Proxima Centauri*, at a distance of 4.243 light years. Deduce this distance in metres and in kilometres, giving your answers in scientific notation.

7 A red photon has 2.00 eV of energy. Express this energy in joules.

Estimating

Estimates are useful when it is either not possible to record accurate measurements, or when an approximate value is all that is needed. Also, when planning an experimental investigation it is useful to make an estimate of the effect a change in the independent variable will have on the dependent variable. This will help you to decide on the size of increments by which you should change your independent variable and also to select the most appropriate apparatus and instruments you will need. For example, if you are expecting that with each change in the resistance of a circuit the current will change by a number of amps, then you would not use a milliammeter.

Estimates are also useful when you are checking calculations. For example, if you were measuring the density of a metal alloy object and obtained a result of 90 kg m⁻³, you would see that an error must have occurred if you knew that the density of water is 1000 kg m⁻³. You would expect the measured density of the alloy to be greater than the density of water. Before carrying out a calculation of a physical quantity, it is helpful to make a mental estimate of the approximate expected result.

> **Expert tip**
>
> Here are some common estimated values that may be useful.
>
> - Density of water = 1000 kg m⁻³
> - Density of air = 1 kg m⁻³
> - Weight of an apple = 1 N
> - Current in a domestic kettle = 13 A
> - emf of a car battery = 12 V

2 Practical techniques

Measuring length

■ Metre rule

A metre rule should allow lengths to be recorded with an uncertainty of ±0.5 mm. The calibration (markings) of a metal rule is likely to be more accurate than a wooden metre rule. Note that it is bad practice to place the zero end of the ruler against one end of the object to be measured and to record the reading at the other end. This may lead to a type of **systematic error** (see Chapter 5) known as a **zero error** (Figure 2.1).

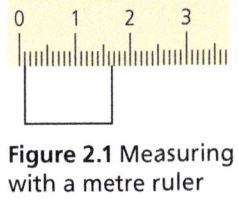

Figure 2.1 Measuring a distance with a metre ruler

You should place the object against the rule so that a reading is recorded at each end of the object. The length of the object is then obtained by subtraction of the two length readings.

The angle at which the scale of the metre rule is viewed will affect the accuracy of the measurement. A non-normal angle of view will give a different reading of the scale (Figure 2.2). This is an example of a **random error** (see Chapter 5) known as parallax error.

> **Key definitions**
>
> **Systematic error** – Experimental error that causes data to be shifted by a consistent amount each time a measurement is made.
>
> **Zero error** – The scale reading of an instrument when the real value is known to be zero. It can be either positive or negative.
>
> **Random error** – Experimental error that causes data to vary in an unpredictable way from one measurement to the next. Readings are spread about a mean value.

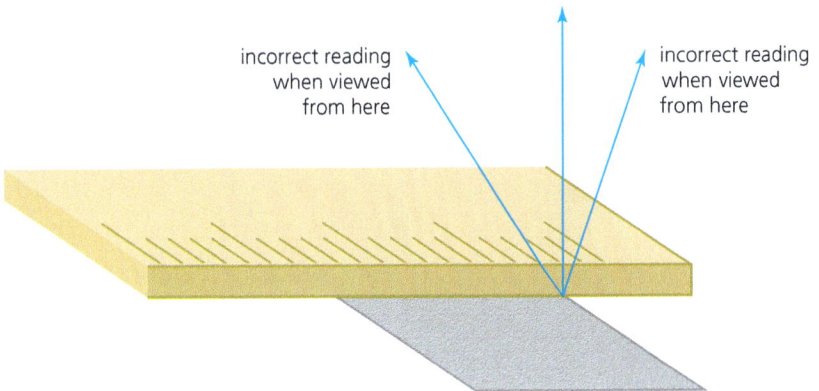

Figure 2.2 Parallax error with a metre rule

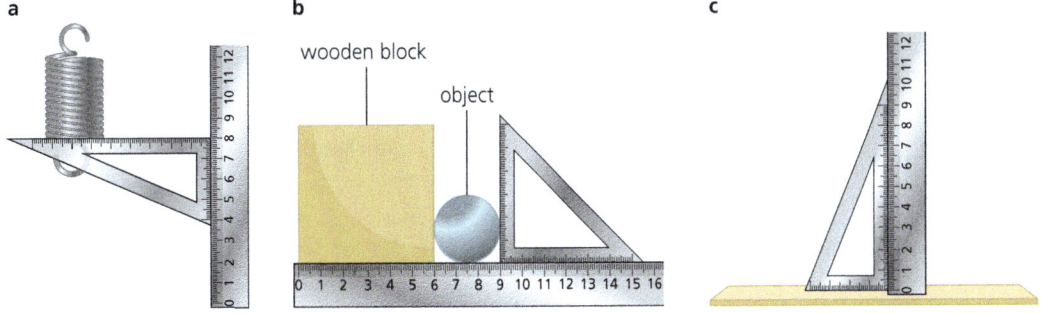

Figure 2.3 Use of a set square to reduce random error

■ Vernier scale

A vernier scale slides along the edge of a main scale, allowing a length to be measured with greater accuracy. n divisions of the vernier scale coincide with $(n-1)$ divisions of the main scale. Lengths can be measured with an accuracy of $\frac{1}{n}$ of the main scale division. Figure 2.4 shows how the scales are read.

> **Expert tip**
>
> A set square can be particularly useful to help give accurate scale measurements by reducing the random error, as shown in Figure 2.3.

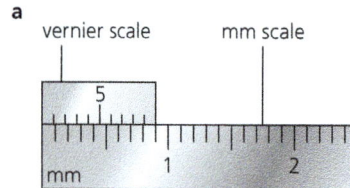

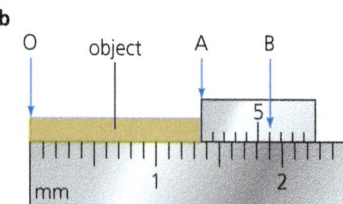

Figure 2.4 Vernier scale: measurement OA = OB − AB = 1.90 cm − (6 vernier divisions) = 1.90 cm − (6 × 0.09 cm) = 1.36 cm

> **Expert tip**
>
> Vernier scales are used in vernier calipers, micrometer screw gauges, barometers, spectrometers and travelling microscopes.

■ Micrometer screw gauge

A micrometer screw gauge (Figure 2.5) is used to measure the dimensions of an object up to about 50 mm (± 10 μm). It relies on the magnification of linear movement using the circular motion of a screw.

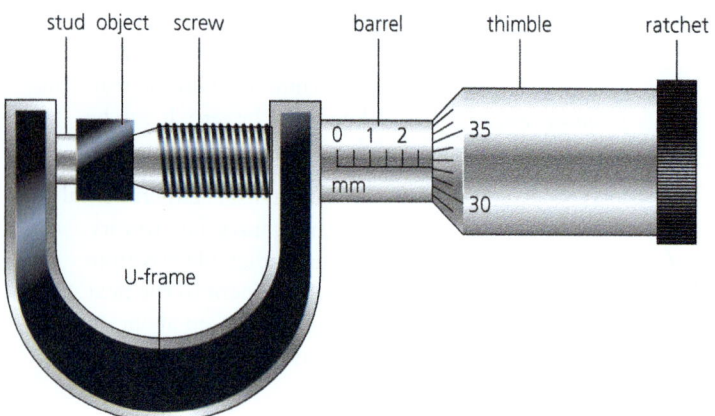

Figure 2.5 Micrometer screw gauge

A micrometer screw gauge consists of a hollow cylinder (the thimble) and a shaft (the barrel) mounted on a U-frame. The U-frame has a flat end (the stud) and a screw on the other side. This screw can be moved by rotating the thimble by means of a ratchet. The object to be measured is placed between the stud and screw. The ratchet is turned to tighten the screw until a 'click' is heard.

The screw moves forward 1 mm for two revolutions of the thimble. The pitch of the screw is therefore 0.5 mm (500 μm). The graduations on the barrel have divisions every 0.5 mm. The reading on the barrel is at the edge of the thimble. When recording a reading, check in which half of the millimetre on the barrel scale the edge of the thimble is. The thimble has a (circular) scale graduated into 50 equal parts. Each division corresponds to $\frac{1}{100}$ of a mm, or 10 μm. Add the reading on the thimble to the reading on the barrel, as shown in Figure 2.6.

> **Expert tip**
>
> The purpose of the ratchet is to ensure that the same torque (turning force) is applied to the thimble for each reading. If this is exceeded, the ratchet slips.

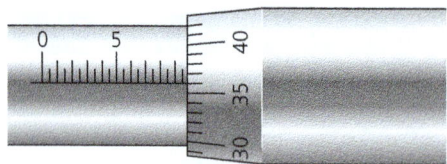

Figure 2.6 Micrometer screw gauge: reading 9.5 mm on the barrel plus 0.36 mm on the thimble, a total of 9.86 mm

Measuring length

Examiner guidance

Check for a zero error having the stud and the screw in contact. The screw must be tightened with the ratchet, so that a reproducible zero is obtained. Record the readings on the barrel and the thimble. For example, Figure 2.7a shows a screw gauge with a zero error of ±0.12 mm. If this was the error when the reading of 9.86 mm was obtained the true length of the object would be 9.86 mm − 0.12 mm = 9.74 mm.

Figure 2.7b shows a zero error of −(0.50 − 0.42) mm = −0.08 mm which means that the true length of the object would be 9.86 mm + 0.08 mm = 9.94 mm.

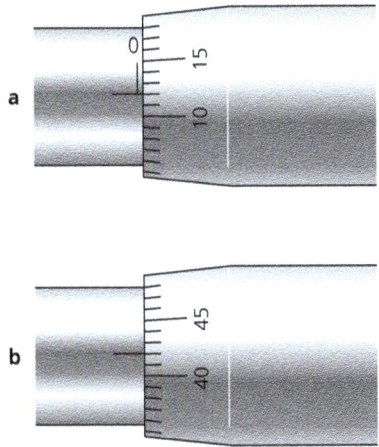

Figure 2.7 Zero error of (a) +0.12 mm and (b) −0.08 mm

■ ACTIVITY

1 In Figure 2.8, parts a, b and c each show a micrometer screw gauge scale, when the jaws are closed and when measuring the thickness of a slab.

   Determine the zero error in each case, and from the observed scale reading work out the actual thickness of the slab.

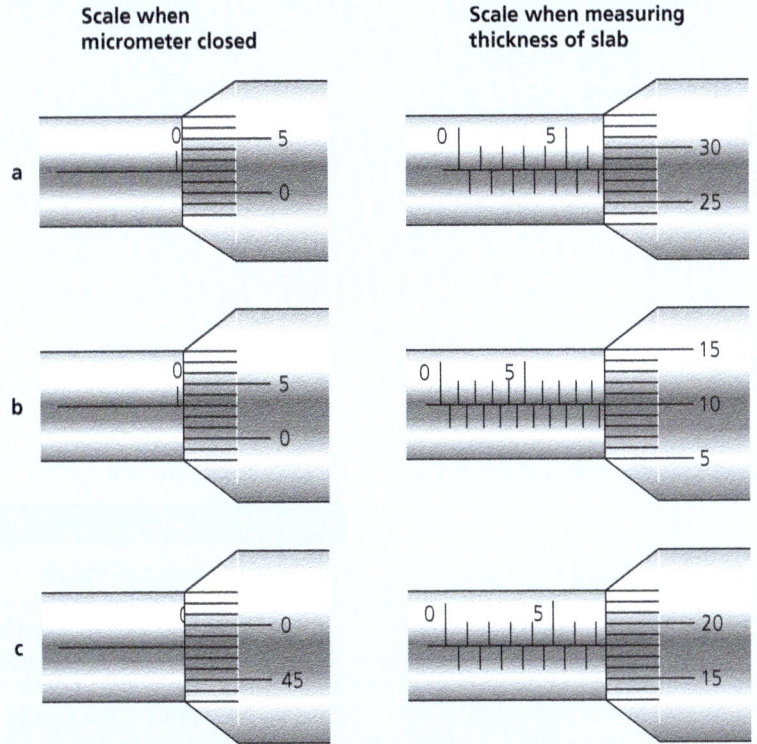

Figure 2.8 Micrometer screw gauge readings activity

RESOURCES

This website gives a description of the use of a micrometer screw gauge in the measurement of Young's modulus of a wire using Searle's apparatus.

https://www.tutorvista.com/physics/searles-apparatus

Vernier caliper

A vernier caliper (Figure 2.9) can be used to measure the dimensions of an object or (using the inside jaws) the diameter of a hole. Its range of measurement is up to about 150 mm, and it can be read to the nearest 0.1 mm or possibly 0.05 mm depending on the type of vernier scale that it has. It consists of a fixed scale and a moveable slider with a vernier scale.

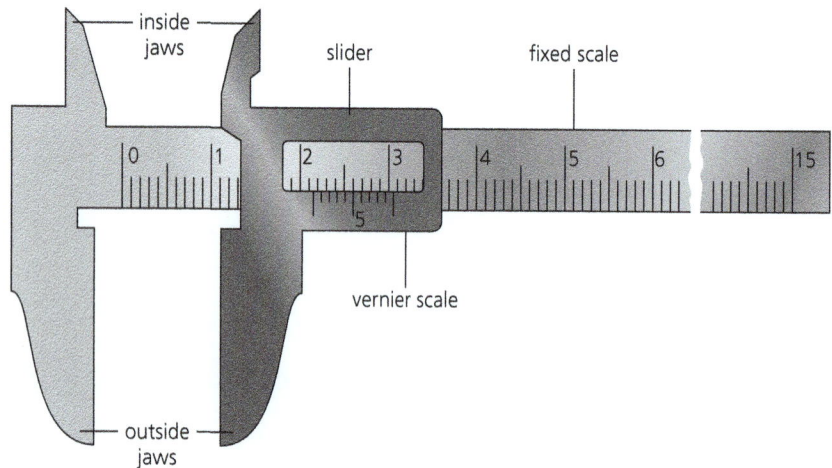

Figure 2.9 The vernier caliper

Figure 2.10 Vernier caliper reading of 26.4 mm

An object is placed between the outside jaws, then the slider is moved along until the object is held tightly. A reading to the nearest millimetre is recorded on the fixed scale, at the zero end of the vernier scale. The reading to a tenth of a millimetre is recorded by observing where a graduation on the vernier scale coincides with a graduation of the fixed scale. Figure 2.10 shows the scale of a vernier caliper giving a reading of 26.4 mm. Zero error readings may be present as described above for the micrometer screw gauge.

The inside jaws have their straight parts on the outside and are used to measure an internal diameter such as that of a hollow cylinder. A pin at the end of the slider can be used to enable the measurement of the depth of a hole.

Worked example

State the reading of the vernier caliper in each of the diagrams.

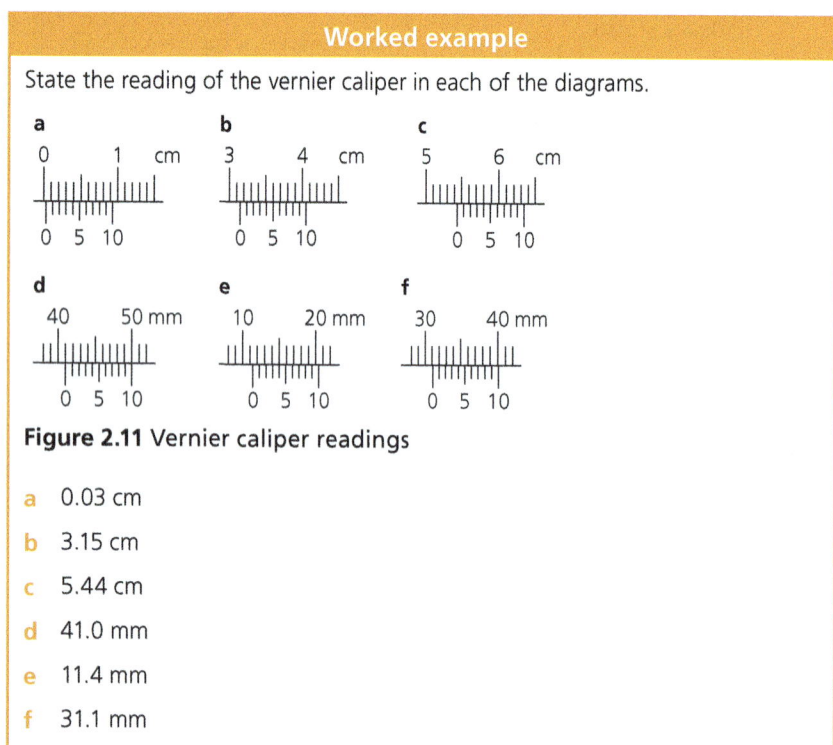

Figure 2.11 Vernier caliper readings

a 0.03 cm
b 3.15 cm
c 5.44 cm
d 41.0 mm
e 11.4 mm
f 31.1 mm

Expert tip

Vernier calipers cannot usually measure objects greater than 15 cm in length, because then the random uncertainty due to thermal expansion of the scales may be greater than the random uncertainty of reading the vernier scale.

If the dimension of a small object is needed, such as the thickness of a microscope slide, and only a metre rule is available, accuracy is improved greatly by measuring the thickness of say 20 identical microscope slides placed side by side. The total thickness is divided by 20 to find the thickness of one slide. Figure 2.12 illustrates this useful physical principle: the accurate measurement of a small quantity by measuring a large number of samples. Even greater accuracy would result if a vernier caliper was used to measure the total thickness of the 20 slides.

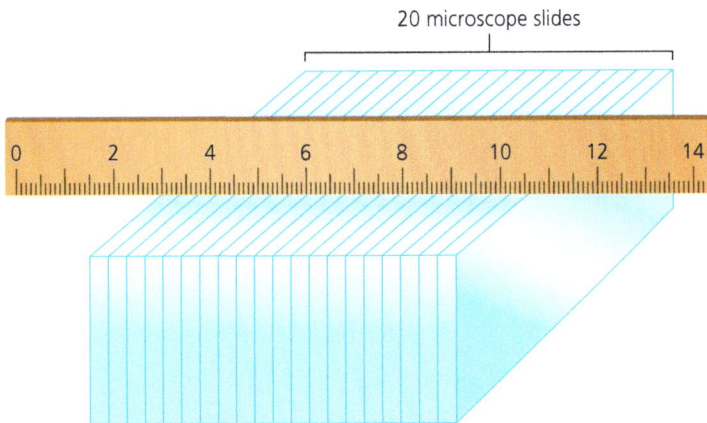

Figure 2.12 To determine the thickness of one glass slide using a metre rule

■ ACTIVITIES

2 400 sheets of paper have a thickness of 2.10 cm. Deduce the average thickness of one sheet of paper in millimetres.

3 In Figure 2.13, parts (a), (b) and (c) each show a vernier caliper scale, when the jaws are closed and when measuring the diameter of a ball bearing.

   Determine the zero error in each case, and from the observed scale reading determine the actual diameter of the ball bearing.

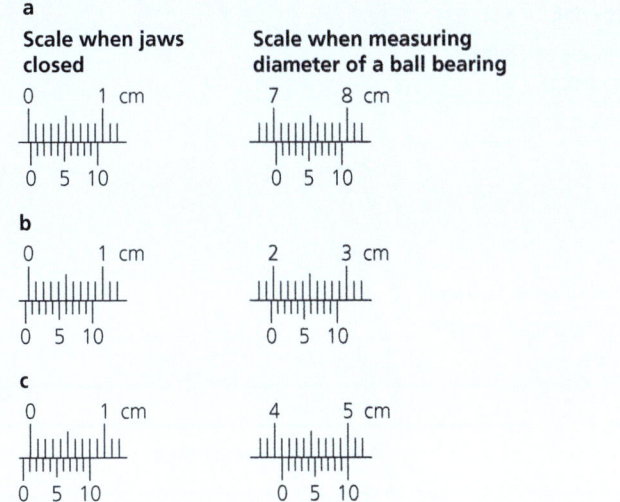

Figure 2.13 Vernier caliper readings

■ Travelling microscope

A travelling microscope (Figure 2.14) is a specialized instrument for accurate length measurement. It consists of a low-power microscope mounted on a slider which is moved along rails (with a fixed scale alongside) by a screw. The microscope is fitted with cross-wires which can be focused on one feature of an object, for example, an edge, and the reading on the fixed scale and vernier (or micrometer) scale recorded. The travelling microscope is then moved along the rails by turning the screw until the cross-wires coincide with the other edge of the object. The difference between the scale readings gives the width of the object.

> **Expert tip**
>
> Uses of the travelling microscope include measuring the separation of interference fringes in a double-slit experiment. The fringe pattern may be displayed on a ground-glass screen and viewed with the travelling microscope, or the fringes may be viewed directly: a pin is placed in front of the microscope to determine the plane of viewing.

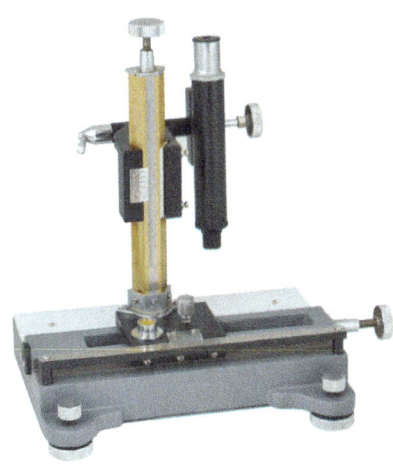

Figure 2.14 A travelling microscope (photo courtesy of Philip Harris)

> **RESOURCES**
>
> This website gives a description of the use of a travelling microscope in the measurement of Young's modulus of a bar:
>
> http://www.nsec.ac.in/images/bes_Young_s_Modulus.pdf

> **Expert tip**
>
> When measuring the diameter of a wire or rod with a micrometer screw gauge, vernier caliper or travelling microscope, it is good technique to record several readings spirally and at intervals along the object's length (Figure 2.15), in case the object has a cross section that is not perfectly circular and in case there is taper (becoming thinner at one end).

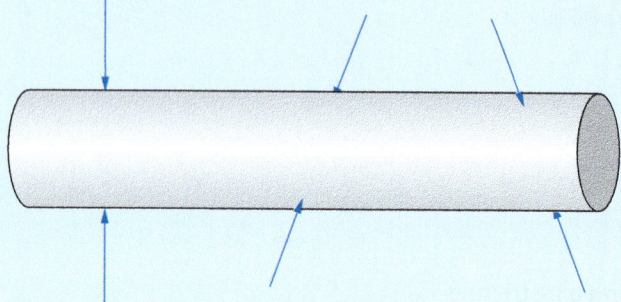

Figure 2.15 Measuring the diameter of a wire

■ Summary of length-measuring instruments

A summary of the range and **resolution** (sensitivity) of length-measuring instruments is given in Table 2.1.

Instrument	Range	Resolution	Notes
Metre rule	1 m	0.5 mm	Check zero, calibration error
Micrometer screw gauge	50 mm	0.01 mm	Check zero error
Vernier caliper	150 mm	0.1 mm	Versatile: inside and outside diameters, depth
Travelling microscope	250 mm	0.1 mm	No contact with object

Table 2.1 Length-measuring instruments

> **Key definition**
>
> **Resolution** – The smallest change in reading that can be detected by a measuring instrument.

■ Area and volume

When carrying out practical work or analysing data from practical work, you may have to calculate different geometric properties of objects with regular or specific shapes. Table 2.2 lists some useful mathematical formulae for this.

Practical activity	Shape	Formula
Area under a force–extension graph for a spring or length of elastic obeying Hooke's law	Triangle	Area = $\frac{1}{2}$ × base × height
Cross-sectional area of a wire	Circle of radius, r	Area = πr^2
Calculating light energy radiated by an incandescent bulb	Sphere of radius, r	Surface area = $4\pi r^2$
Calculating the light energy radiated by a fluorescent tube	Cylinder of radius, r, and height, h	Surface area = $2\pi r^2 + 2\pi r h$
Calculating the density of a rectangular solid block	Rectangular block of sides a, b and c	Volume = abc
Calculating the density of a steel ball bearing	Sphere of radius, r	Volume = $\frac{4}{3}\pi r^3$
Calculating the volume of a gas syringe	Cylinder of radius, r, and height, h	Volume = $\pi r^2 h$

Table 2.2 Some useful mathematical formulae for geometric properties

Worked example

A vertical uniform tube contains a column of liquid of length 12.0 cm and a base area of 2.0 cm², as shown in Figure 2.16.

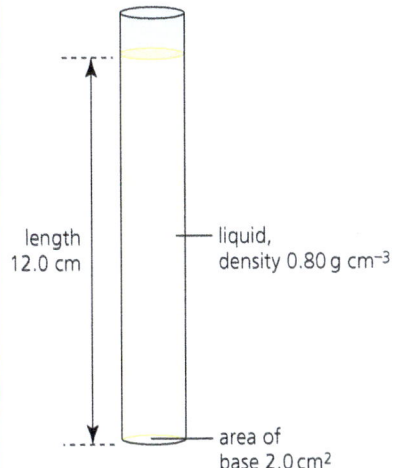

Figure 2.16 Vertical uniform tube containing liquid

Calculate the mass of the liquid in the tube.

Volume of liquid = 2.0 cm² × 12.0 cm = 24.0 cm³

Mass of liquid = 24.0 cm³ × 0.80 g cm⁻³ = 19.2 g

Measuring angles

Angles are measured in degrees using a protractor. To measure the angle between two lines, the centre of the protractor is placed exactly over the point where the two lines meet (intersect) and one line is aligned with the zero degree direction of the protractor. The angle between the lines is then given by the reading on the scale at which the second line passes through the circumference of the protractor's circle or half circle.

With ordinary school protractors it is easy to record a reading to the nearest degree, and sometimes to half a degree if an angle between fine lines is being measured. Measuring to the nearest 0.5° requires **interpolation** between the divisions on the scale.

Expert tip

If you can measure an angle with a protractor to 0.5° then for an angle of 5.0° this would give rise to a **percentage uncertainty** of ±10%. By using trigonometric methods to measure small angles, you can significantly reduce this uncertainty and hence improve accuracy. If you can measure the adjacent and opposite sides using a ruler, as shown in Figure 2.17, then the angle can be calculated from:

$\theta = \tan^{-1}\left(\dfrac{\text{opposite}}{\text{adjacent}}\right) = \tan^{-1}\left(\dfrac{h}{l}\right)$.

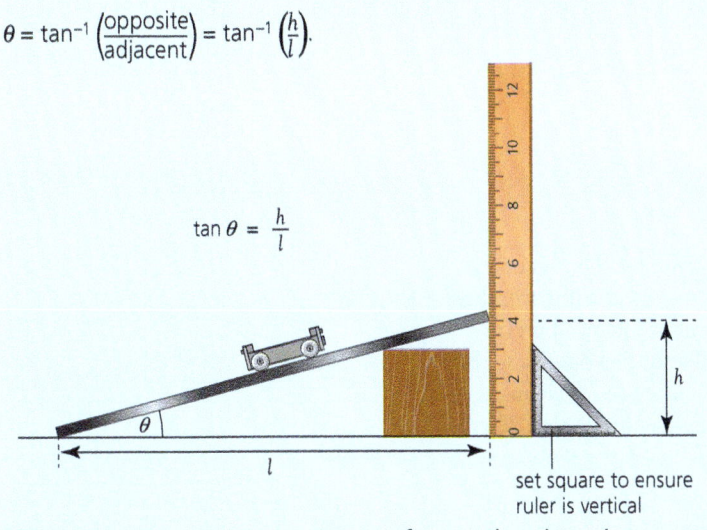

Figure 2.17 Accurate measurement of an angle using trigonometry

Key definitions

Interpolation – Estimation of a value for a variable between two or more known values, or of a scale reading between two scale graduations.

Percentage uncertainty (percentage error) – An uncertainty or error in a measured value, expressed as a percentage of the value.

■ ACTIVITY

4 Calculate the angle of incline made by a 1.00 m ramp to the horizontal if the height of the raised end is 85 mm above the bench. Estimate the percentage uncertainty in the value for the angle.

Measuring mass

Masses are measured using a balance. Balances compare the weight of the unknown mass with the weight of a standard mass. Since weight is **directly proportional** to mass, the unknown mass is being compared with the standard mass. Your physics laboratory may have a top pan (electronic) balance (Figure 2.18), a spring balance or a lever balance.

> **Key definition**
>
> **Directly proportional** – As one variable increases, the other increases by the same percentage.

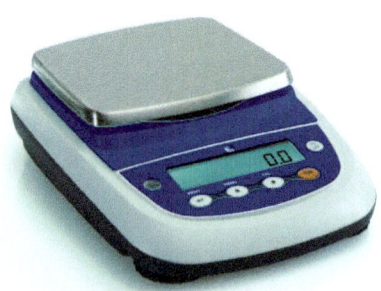

Figure 2.18 A top pan balance (source: Edulab)

Spring balances (also called force meters or newton meters) are based on Hooke's law: the extension of a loaded spring is directly proportional to the weight of the load (provided the elastic limit is not exceeded). The extension is measured directly, by a marker moving along a straight scale, or by a pointer moving over a circular scale.

Lever balances are based on the principle of moments. The unknown mass is placed on a pan, and balance is achieved by sliding a mass along a bar, calibrated in units of mass, until the bar is horizontal. The moment of the load is equal and opposite to the moment of the sliding mass and the bar. A reading is taken from the edge of the sliding mass on the divisions marked on the bar. Alternatively, a lever balance may have a pointer that moves along a circular scale (Figure 2.19).

> **Expert tip**
>
> Some types of spring balance are calibrated in newtons (force units), rather than in kilograms (mass units).

Figure 2.19 A lever balance (source: Edulab)

Measuring volume

The volume of a liquid may be found using a measuring cylinder. The accuracy depends on the size of the measuring cylinder; for a 100 cm^3 measuring cylinder, the accuracy is to the nearest cubic centimetre. For greater accuracy a burette may be used. Burettes can contain up to 50 cm^3 of liquid and the volume can be measured accurately to 0.10 cm^3. Pipettes are used to dispense accurately measured small volumes of liquid; they are available in various sizes, usually 10 cm^3, 20 cm^3, 25 cm^3, and 50 cm^3.

Measuring volume

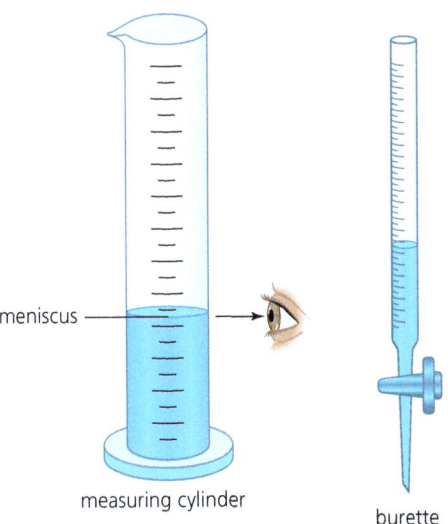

Figure 2.20 A measuring cylinder and a burette

> **Expert tip**
>
> When reading the water level in a measuring cylinder or burette to find the volume, the eye must be on the same level as the bottom of the water meniscus to avoid parallax error.

Volumes of regular solids are found by taking accurate measurements of their dimensions and using formulae to determine the volume. Volumes of irregular solids are measured by a displacement method, illustrated in Figure 2.21. Water is poured into a measuring cylinder and the volume of water is recorded. An object is placed in the water so that it is fully submerged and the water is displaced. The new volume is recorded; the volume of the object is equal to the difference in reading for the two volumes of water.

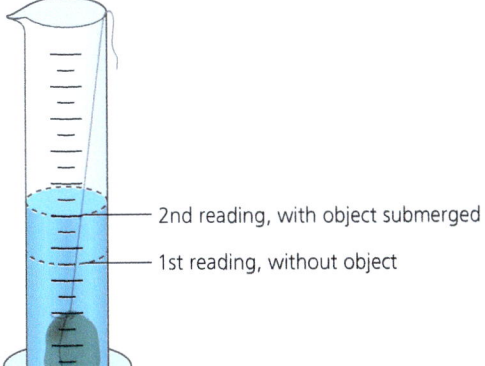

Figure 2.21 Measurement of the volume of an irregular solid

An alternative method for finding the volume of irregular solids uses the overflow (or Eureka) can (Figure 2.22). The can is filled with water and allowed to stand until it no longer drips; the object is then placed in the can and the displaced water collected in a beaker or measuring cylinder; the volume of the displaced water is equal to the volume of the object.

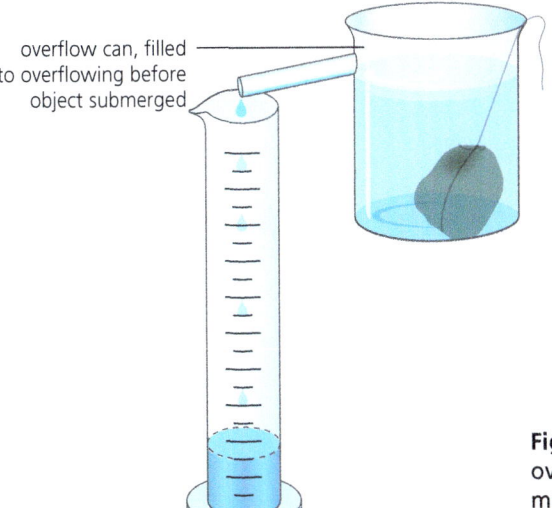

Figure 2.22 Using an overflow (Eureka) can to measure the volume of an irregular solid

Measuring time

An electronic stopwatch has a digital display and its operation is based on the oscillation of a quartz crystal. The readout is usually to the nearest one-hundredth of a second (0.01 s).

> **Examiner guidance**
>
> In experiments where the stopwatch is started and stopped by hand (manually), a precision of 0.01 s is not justified. Your reaction time (typically 0.3 s) will cancel out the precision of the stopwatch. For example, an electronic stopwatch may display 22.112 s, but this might be best reported as 22.1 s ± 0.3 s.

A light gate is a digital sensor which consists of an infrared transmitter and receiver mounted in a strong frame with a small gate gap. It can be used to record the starting time, finishing time and hence duration of an event.

> **Examiner guidance**
>
> Measurements recorded with electronic sensors are more precise and more accurate than measurements recorded manually using stopwatches. The electronic measurements do not suffer from errors such as reaction time or possible misreading of scales (if an analogue stopwatch is used). This means that the results are more reliable.

Measuring movement

An accurate measurement of movement is needed in order to measure the effect of a force. In dynamics experiments this is best done using a ticker-timer (or ticker-tape timer), shown in Figure 2.23. Test the ticker-timer by closing the switch and pulling the paper tape through the timer; a series of dots should be marked on the tape, each one made by the vibrating marker hitting the tape.

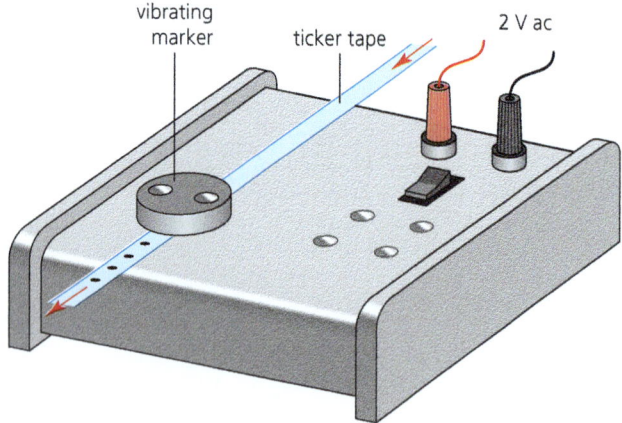

Figure 2.23 Ticker-tape timer

> **Expert tip**
>
> Ticker-timers usually vibrate at exactly 50 ticks per second (50 Hz). Each dot on the ticker tape occurs exactly 0.02 s after the one before.

The distance between dots on a ticker tape represents the object's position change during that time interval. A large distance between dots indicates that the object was moving fast during that time interval. A small distance between dots means the object was moving slowly during that time interval.

The analysis of a ticker-tape diagram (Figure 2.24) made of 'tentick lengths' will also show if the object is moving with a constant velocity or accelerating. A changing distance between dots indicates a changing velocity and thus an acceleration. A constant distance between dots represents a constant velocity and therefore no acceleration.

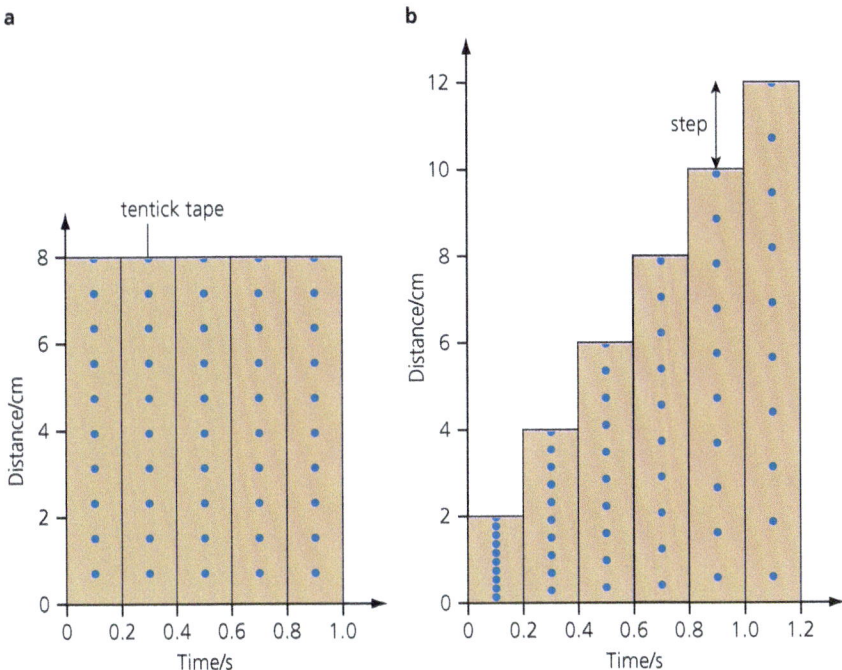

Figure 2.24 Tape charts for **a** uniform speed and **b** uniform acceleration (note the steps are of equal height)

Measuring temperature

■ Liquid-in-glass thermometers

Liquid-in-glass thermometers use the thermal expansion of a liquid (mercury or coloured ethanol) contained in a bulb at the end of a thin capillary tube. The space above the liquid contains an inert gas at low pressure (usually nitrogen). The most common thermometer is a mercury-in-glass thermometer (Figure 2.25) with graduations from –10 °C to 110 °C, in 1 °C intervals.

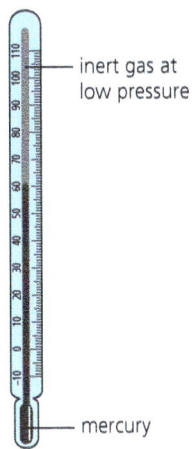

Figure 2.25 Mercury-in-glass thermometer

> **Expert tip**
>
> Allow time for the thermometer to reach thermal equilibrium before recording the temperature. Any liquids must be stirred before a temperature reading is taken, so the heat is distributed evenly in the liquid. Because of convection currents there can be a temperature difference between the top and the bottom of a liquid.

■ Thermocouple thermometer

A thermocouple consists of wires of two different metals, such as copper and iron, joined together (Figure 2.26). When one junction is at a higher temperature than the other, an electric current flows. The corresponding voltage or emf can be measured by a digital voltmeter set to measure millivolts. One junction is placed in thermal contact with the object whose temperature is being measured. The voltmeter can be **calibrated** so that a given voltage will correspond to a temperature in degrees Celsius.

> **Key definition**
>
> **Calibrate** – Align a measuring instrument's scale with known points or values.

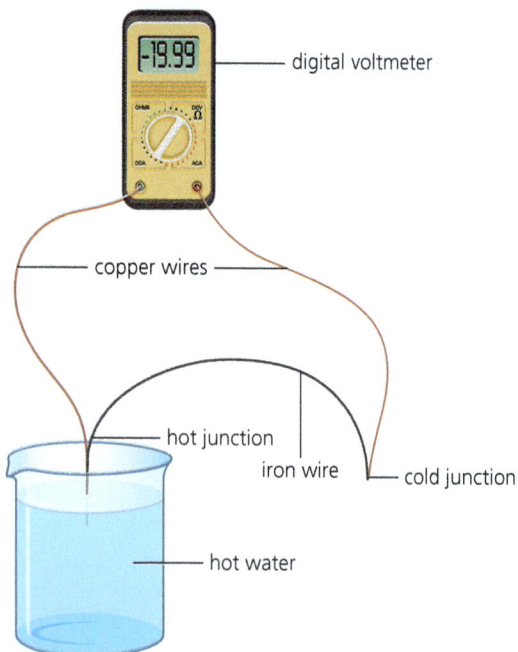

Figure 2.26 A simple thermocouple thermometer

A thermocouple thermometer actually measures the *difference* in temperature between the hot and cold junctions of the two metals. If the voltmeter has not been calibrated in degrees Celsius then a calibration curve needs to be drawn.

Investigations

- Calibrate a thermocouple.
- Investigate factors that affect the time for a liquid to totally evaporate, with a focus on the Leidenfrost effect. This is a physical phenomenon in which a liquid, in near contact with a mass significantly hotter than the liquid's boiling point, produces an insulating layer that keeps the liquid from boiling rapidly.

Examiner guidance

- The heat capacity of the hot junction of a thermocouple is much smaller than that of the bulb of a liquid-in-glass-thermometer. Hence, the thermocouple is especially useful if a varying temperature is to be measured, or if the object whose temperature is to be measured has a small heat capacity.
- A thermocouple thermometer can measure a much wider range of temperatures than a mercury-in-glass thermometer.
- A thermocouple thermometer can measure temperature at a point (because the junction is very small).

■ ACTIVITY

5 Figure 2.27 shows the calibration curve for a thermocouple used to measure temperatures from 0 °C to 250 °C. Use interpolation to deduce the value of the emf of the thermocouple at a temperature of 235 °C.

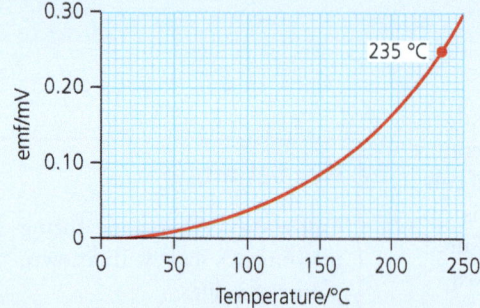

Figure 2.27 Calibration curve for a thermocouple

Measuring gas pressure

A difference in gas pressure may be measured by comparing the heights of liquid in the two arms of a U-tube. Figure 2.28 shows a U-tube connected to a container of gas. The pressure above the liquid in tube B, and hence the pressure in the container, is p.

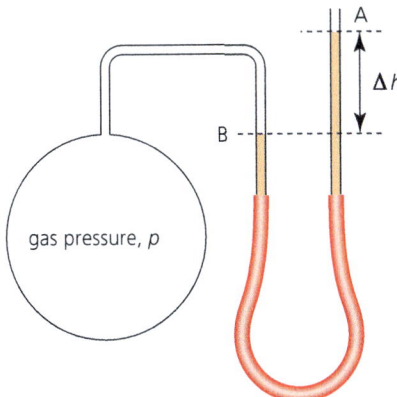

Figure 2.28 A U-tube with liquid connected to a container of gas

$p = p_0 \pm \rho g \Delta h$

where p_0 is the (atmospheric) pressure at A, Δh is the difference in vertical height between the levels of the liquid in the two arms of the tube, ρ is the density of the liquid and g is the acceleration of free-fall. To find the pressure of the gas, p, you need to measure the difference Δh between the liquid levels, assuming the other variables are known.

A U-tube mounted on a board to which a vertical millimetre scale is attached is known as a manometer and it usually contains oil or water. A set square can be used to find the reading on the vertical scale corresponding to the centre of the meniscus in each side of the U-tube (Figure 2.29).

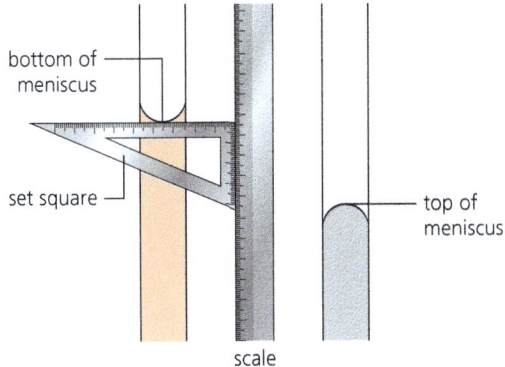

Figure 2.29 Reading the level of the liquid in a manometer

> **Expert tip**
>
> If the manometer contains mercury the meniscus will be convex and the measurement is to the top of the meniscus.

> **RESOURCES**
>
> See this website for the determination of surface pressure by a bubble pressure method.
>
> www.fpharm.uniba.sk/fileadmin/faf/Pracoviska-subory/KFCHL/ENG/lectures/Physics/2.buble_pressure.pdf

Measuring current and potential difference

The two main types of ammeter and voltmeter are analogue meters (a pointer moves over a scale) and digital meters.

Since current is the rate of flow of charge past a single point, in order to measure the current it is necessary to break into the circuit at that point and connect the current-measuring device so that the charge flows through it.

Potential difference (pd) is measured *between* two different points in a circuit. To measure the pd between two points, two probes from the measuring device need to be connected across the two points (that is, in parallel with components through which current is flowing).

Analogue meters

The scales of analogue meters are restricted to the measurement of currents or potential differences over a single range, for example, a 0–1 A dc ammeter will measure direct currents in the range from 0 to 1 A.

Analogue meters may have a zero error (Figure 2.30). Before switching on the electric circuit, check whether the needle is exactly at the zero mark. If it is not, move the needle (pointer) by adjusting the screw at the needle pivot. Read the needle position from directly above it and the scale, and not from one side, to avoid parallax error.

> **Expert tip**
>
> Some analogue meters have a dual-range facility, with a common negative terminal and two positive terminals, each of which is associated with an upper or lower scale on the ammeter or voltmeter. Each of the positive terminals is marked with the scale to which it refers. Make sure you record the reading on the scale corresponding to the pair of terminals you have selected.

> **Expert tip**
>
> When measuring sinusoidal alternating values, remember that the reading obtained on an analogue or digital voltmeter is the rms value.

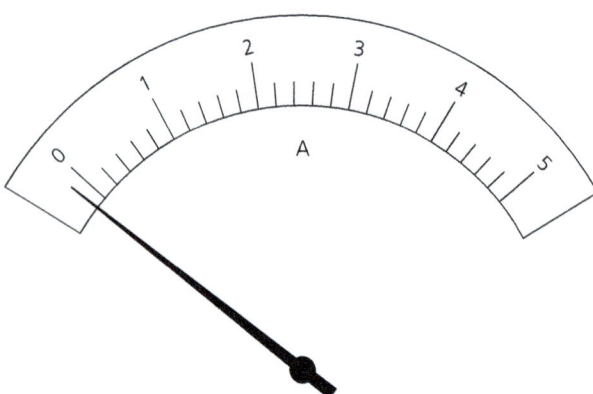

Figure 2.30 A zero error (of about −0.2 A) on the analogue scale of an ammeter

Galvanometers

A galvanometer is a sensitive analogue meter which may be converted into an ammeter by the connection of a suitable resistor (known as a shunt) in parallel with the meter. The meter may be converted into a voltmeter by the connection of a suitable resistor (known as a multiplier) in series with the meter (Figure 2.31).

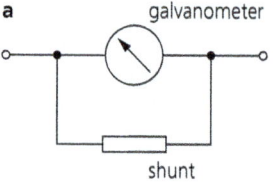
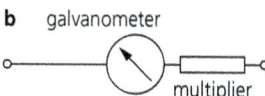

Figure 2.31 a Galvanometer with shunt, for current measurements; **b** galvanometer with multiplier, for voltage measurements

> **Expert tip**
>
> The heating effect causes significant changes in the electrical resistance in metallic conductors. There are various ways of minimizing the heating effect:
>
> - Keep the current relatively small and turn off when you are not recording a measurement.
> - Ensure adequate ventilation in the lab or use a cooling fan.
> - Use a liquid for cooling, for example, immerse a resistance wire in cold water.
> - Use physically large components, or combinations to share the heating effect (Figure 2.32).
>
>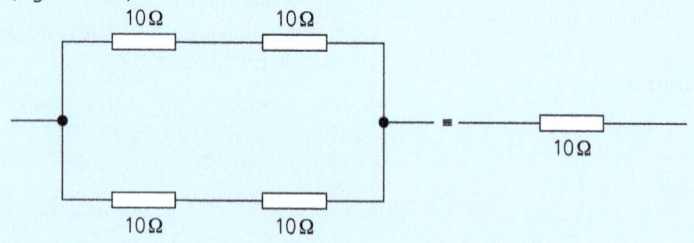
>
> **Figure 2.32** Reduction of heating effect in resistors

Digital meters

Manufacturers of digital meters quote the uncertainty for each meter, for example, ±1% ± 2 digits. The ±1% applies to the total reading on the scale and the ±2 digits is the uncertainty in the final figure of the digital display. This means that the uncertainty in a reading of 3.00 V would be $\left(\pm 3.00 \times \frac{1}{100}\right) \pm 0.02 = 0.05$ V. This inherent uncertainty would be added to any further uncertainty due to a fluctuating reading.

Hot-wire ammeter

The hot-wire ammeter is a historical device, also known as the thermal ammeter, that may be a suitable instrument to investigate (see Chapter 7). It is based on the heating effect of an electric current and may be used for measuring ac as well as dc. Instruments used for measuring ac need to be designed so that the pointer deflects the same way when the current flows through the instrument in either direction.

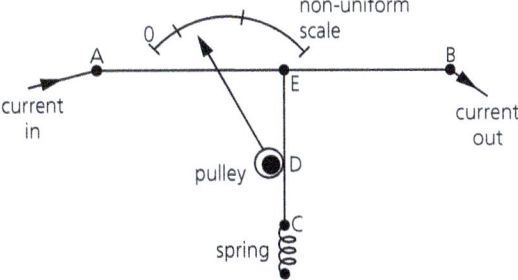

Figure 2.33 Principle of one type of hot-wire ammeter

In the hot-wire ammeter shown in Figure 2.33, heat is produced in the thin resistance wire AB when current (I) flows. When the temperature of wire AB increases, it expands and sags. The slackness is taken up by another fine wire DC, which passes round a pulley and is kept tight by a spring. Any movement of the wire DC causes the pulley to rotate, so the pointer moves over the scale. The deflection θ of the pointer is approximately proportional to the average rate at which thermal energy is produced (that is the power) in the wire AB.

$\theta \propto P$ and $P \propto I^2$ (assuming resistance, R, is constant), so $\theta \propto I^2$

Hence the deflection of the pointer is approximately proportional to the square of the current and so the scale is not uniform. In ac measurements the meter reads the rms current if sinusoidal.

Multimeters

It is very common for a single instrument to incorporate both a voltmeter and an ammeter. It may also be able to perform other types of electrical measurements, such as resistance. A device that combines several different electrical measurement functions is called a multimeter (Figure 2.34).

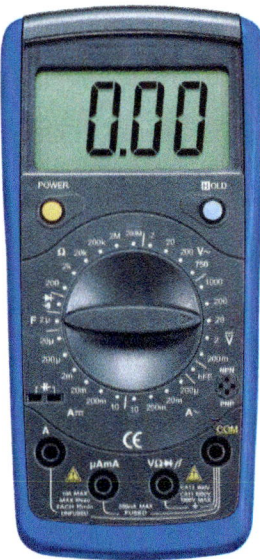

Figure 2.34 Digital multimeter (source: Edulab)

You can select the function and the measurement range by rotating the central dial. Each range setting is labelled with the maximum value measurable on that setting.

> **Expert tip**
>
> Use the lowest range setting that is greater than the size of the maximum measurement that you expect to measure. If you try to measure a value that exceeds the maximum for that range, the display screen of the multimeter will either flash or display 1. You must then increase the range setting until you see a sensible reading on the display.

> **Investigations**
>
> A digital multimeter may be used to:
> - verify the formulae for resistors in series and in parallel
> - investigate how the resistance of an LDR varies as the intensity of the incident light varies
> - investigate how the resistance of a thermistor varies as its temperature changes
> - measure body/skin resistance.

Oscilloscopes

The cathode ray oscilloscope (CRO) is a very useful device used for displaying and measuring electrical signals, in particular those varying too rapidly to be measured with a meter. A traditional analogue oscilloscope has a cathode ray tube, where electrons emitted by a heated filament are focused into a beam and accelerated by a high voltage until they strike a phosphor screen, transferring their kinetic energy and producing a spot of light.

Voltages applied to pairs of metal plates in an analogue oscilloscope can deflect the beam either horizontally or vertically, so moving the spot on the screen.

The main use of an oscilloscope is to display the variation of a voltage signal with time. The vertical axis of the screen displays the voltage being investigated while the horizontal axis shows time. The scale of the two axes is set by the 'gain' (or 'Y-gain') and the 'time base' controls respectively. In order to obtain a steady-state trace for a repetitive signal, the time base may be synchronized to one of the input signals. This is done via the 'trigger' controls.

Another mode of operation, known as 'X-Y display', displays the voltage applied to one input as a function of the voltage applied to the second input.

Figure 2.35 shows a traditional two-channel (dual trace) laboratory CRO.

Expert tip

A digital meter has a response time, which is the time taken to give a reading after a change in input. If the response time is relatively long compared with the time taken for a change of input, the meter will not measure changing inputs reliably.

Expert tip

Modern digital oscilloscopes use smaller liquid crystal display screens that display a waveform in response to the time-variant voltage input to it. Such a design effectively emulates the traditional analogue oscilloscope.

Expert tip

There are two sources of error in using an oscilloscope: the calibration error and the reading error. The calibration error is related to the accuracy of the components used within the oscilloscope and is assumed to be 5% for both voltages and times. The reading error results from the precision with which a reading can be made from the screen, and is often related to the line width of the trace. The total error of any reading is the combination of the calibration and reading errors.

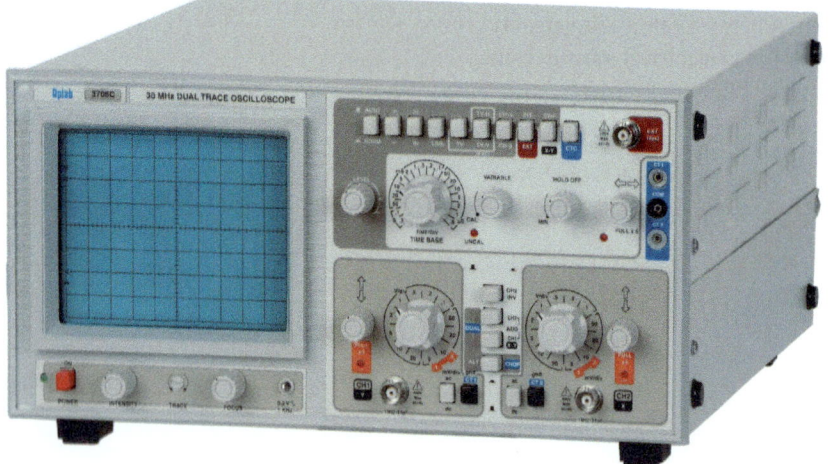

Figure 2.35 A two-channel laboratory CRO (photo courtesy of Philip Harris)

■ Measurement of voltage using a CRO

A potential difference applied to the Y-input controls the movement of the trace in a vertical direction. A pd applied across the X-input controls the trace in the horizontal direction. The Y-scale sensitivity (Y-gain) is adjustable and is measured in volts per centimetre (V cm^{-1}) or volts per division (V div^{-1}).

Suppose that the Y-gain is set at 2 V div^{-1}. A dc power supply is to be connected across the Y-input, but no voltage is applied across the X-input. The trace appears as a bright spot.

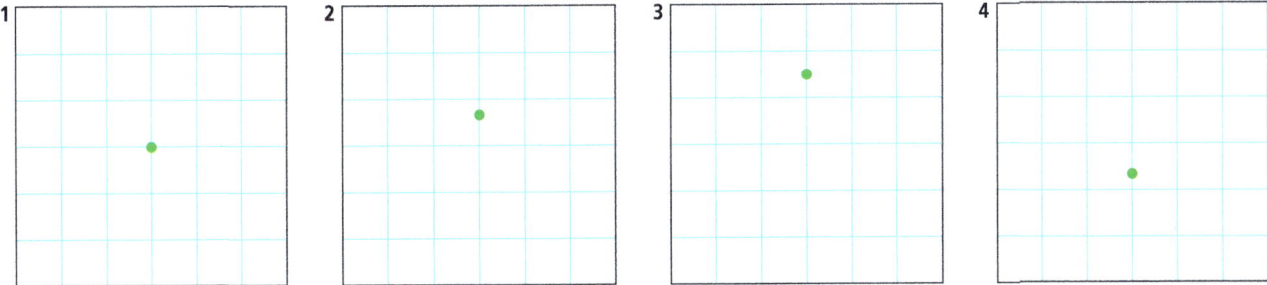

Figure 2.36 Using an oscilloscope to measure potential difference (voltage)

In Figure 2.36, screen 1 shows the CRO screen with no input voltage. Screen 2 shows a deflection of +0.75 of a division; the voltage input is therefore +0.75 × 2 Vt = +1.5 V. Screen 3 shows a deflection of 1.5 divisions; the voltage input across the Y plates is +1.5 × 2 V = +3.0 V. Screen 4 shows a deflection of −0.75 divisions; the voltage input across the Y plates is −0.75 × 2 V = −1.5 V.

> **Examiner guidance**
>
> High input resistance devices, such as oscilloscopes and digital voltmeters, can significantly increase accuracy of measured voltages.

Measurement of time intervals using a CRO

To measure short time intervals, a repeating time base voltage is applied across the X-input (Figure 2.37). This moves the spot across the screen, before returning to the start again. The rate at which the time base voltage drags the spot across the screen can be measured in either seconds per division (s div^{-1}) or as a frequency in divisions per second (div s^{-1}).

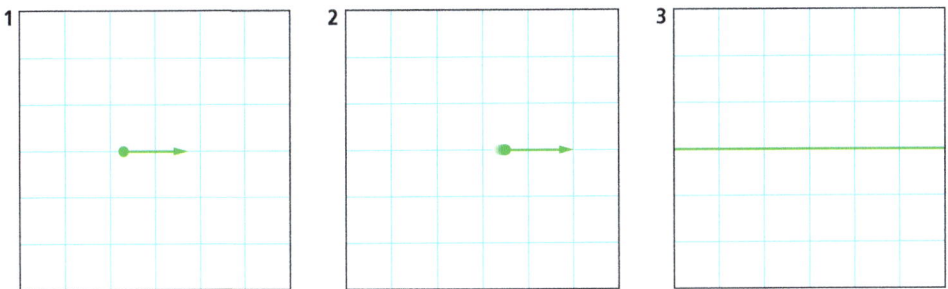

Figure 2.37 Using an oscilloscope to measure a time interval

Screen 1 of Figure 2.37 shows the spot moving slowly across the screen before returning to the start where the process repeats. In screen 2 the spot moves more quickly because of a higher frequency time base voltage, and a short tail is formed. In screen 3, with a much higher time base frequency, the fluorescence of the spot lasts long enough to form a continuous line, or trace.

> **Investigations**
>
> You can use an oscilloscope (or data logger and computer) to:
>
> - display and analyse the pd across a capacitor as it charges and discharges through a resistor
> - determine the speed of sound in air (using a dual trace oscilloscope, signal generator, speaker and microphone)
> - investigate the effect on magnetic flux linkage of varying the angle between a search coil and a magnetic field.

3 Using apparatus

Using an electroscope

The gold leaf electroscope (Figure 3.1) consists of an isolated conductor (the metal rod, which is insulated from the ground) with a gold leaf attached to indicate the presence of excess charge, of either sign. The leaf rises because it is repelled by the rod and plate when they have excess charge of the same type.

The gold leaf electroscope is a simple, sensitive and semi-quantitative instrument. It can be used to measure the pd between the gold leaf and the case (or earth).

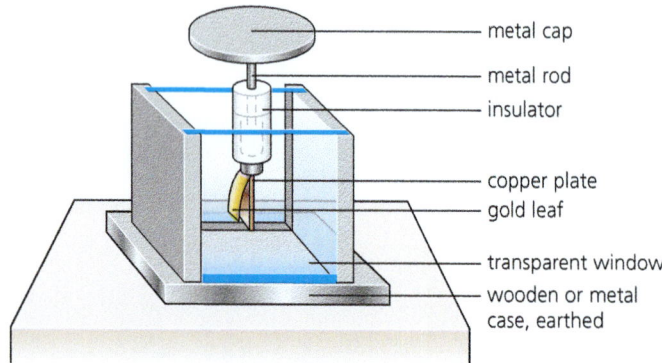

Figure 3.1 A gold leaf electroscope

> **Examiner guidance**
>
> A risk assessment should be performed for all the apparatus and instruments you plan to use and actually use. There must be no flammable vapours during experiments involving static and current electricity. An ohmmeter is an electrical instrument that is more of a risk than the gold leaf electroscope, because a current flows when it is used. You should avoid holding or touching the length of wire whose resistance is being measured. Rubber-soled shoes should be worn at all times in the lab to minimize any risk of electrocution. Water and samples of conducting metals not connected with the experiment must be removed from the bench.

■ Charging an electroscope
■ Charging by contact and induction

An electroscope can be charged by contact (Figure 3.2a). A dry insulator is rubbed to generate an electrostatic charge. It is then touched on the metal cap of the electroscope, which will conduct charge from the insulator to the rod, plate and gold leaf. The charge transferred in this way may not be enough to produce a large deflection.

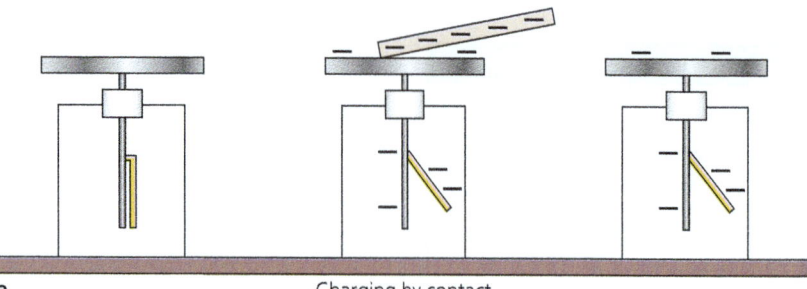

a   Charging by contact

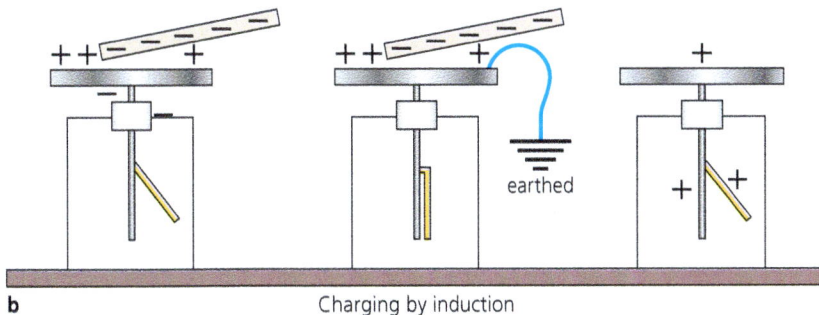

b   Charging by induction

Figure 3.2 Charging a gold leaf electroscope by contact and induction

An electroscope can also be charged by induction (Figure 3.2b). A dry insulator is rubbed to charge it and brought close to the metal cap. The particles that move are electrons, so negative charge is repelled away from the cap, leaving an excess of positive charge on the cap. The gold leaf will be repelled and rise. If the cap is touched momentarily while the charged insulator is still held near it, it will be earthed. Excess charge of the whole system of charged insulator and electroscope will be neutralized. Then the charged insulator is removed. The charge remaining on the cap will now redistribute, leaving the device charged – but with the *opposite* charge to that on the charged insulator. The gold leaf will probably show greater deflection than when charged by contact.

> ### Investigations
> Project a shadow of the gold leaf onto a screen so that the angle of deflection can be measured with a protractor. Connect the electroscope to a high voltage dc supply. Connect one terminal to the cap using a crocodile clip, and the other to the outer casing (there is usually an earthing terminal). Obtain readings and plot angle against pd. The calibrated electroscope (an 'electrometer') could then be used to measure induced voltages from static charges.

> ### Examiner guidance
> The van de Graaff generator is another device that may be used, under teacher supervision and instruction in the laboratory, to investigate static electricity. The generator produces a continuous supply of charge on a large metal dome when a rubber belt is driven by an electric motor (or by hand), as shown in Figure 3.3. It must not be charged by EHT ('extra high tension').

> ### RESOURCES
> These websites describe investigations of the photoelectric effect (the release of electrons from a metal surface in the presence of ultraviolet light) using a gold leaf electroscope:
> - https://www.stem.org.uk/resources/elibrary/resource/28841/photoelectric-effect
>
> Investigations of the photoelectric effect: release of electrons from metal surface in the presence of ultraviolet radiation:
> - https://www.cta-observatory.ac.uk/wp-content/uploads/2013/05/Photoelectric-Effect-teachers-guide.pdf

Using a Hall probe

The Hall effect (Figure 3.4) refers to the potential difference that builds up across opposite faces of a thin cuboidal sample of semiconductor material when it is carrying a current and in a magnetic field. The effect is greatest when the magnetic field is applied in a direction perpendicular to that of the flow of current.

A Hall probe makes use of this effect to measure the magnetic flux density of a magnetic field. It is positioned so that its thin slice of semiconductor material has its plane at right angles to the direction of the magnetic field. The control unit passes a current of known value through the semiconductor slice. The Hall voltage is proportional to the magnetic flux density and is read off a meter (digital or analogue) that has been calibrated in units of magnetic flux density (tesla).

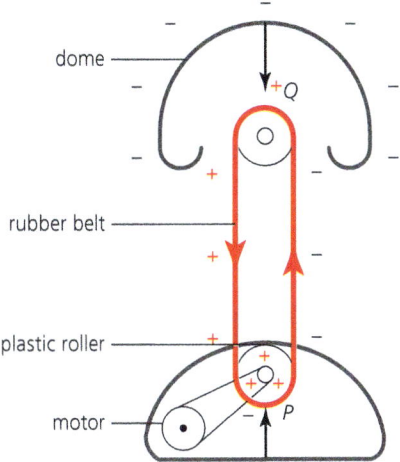

Figure 3.3 Principle of the van de Graaff generator (*P* and *Q* are 'combs' of metal points)

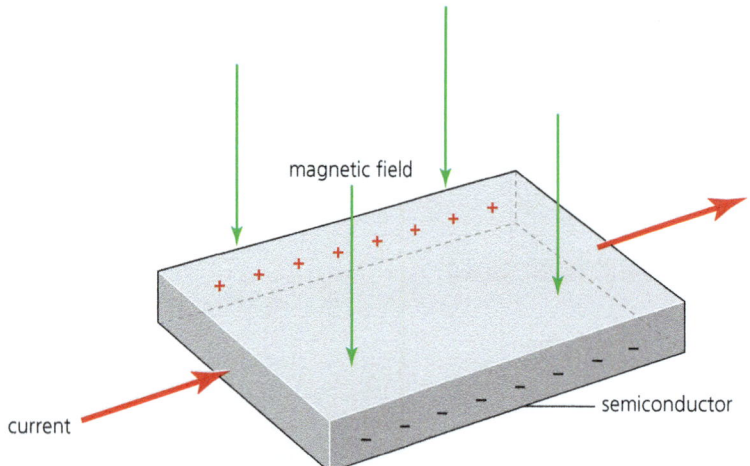

Figure 3.4 The Hall effect

> **RESOURCES**
>
> You can find investigations of the Hall effect at these websites.
>
> - http://www.a-levelphysicstutor.com/field-magnet-2.php
> - https://www.youtube.com/watch?v=ghFNDnd8DHU
> - http://www.schoolphysics.co.uk/age16-19/Electronics/Semiconductors/text/Hall_effect/index.html

The Hall probe can be used to investigate the variation in magnetic flux density with angle, using the magnetic field from a permanent magnet, as shown in Figure 3.5. It can be rotated by varying angles in the field, and the corresponding Hall voltage recorded.

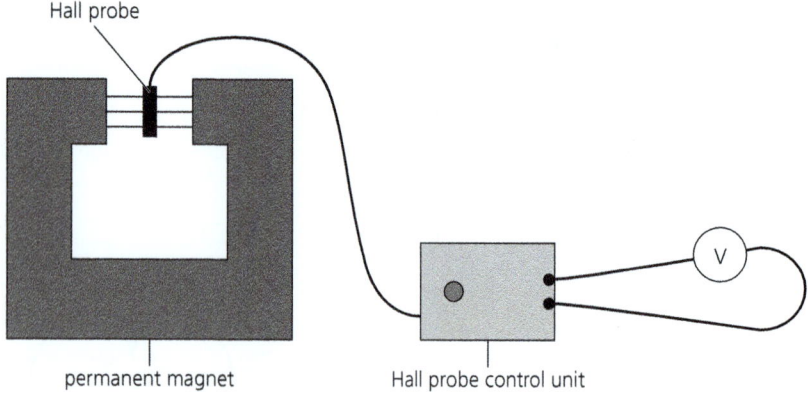

Figure 3.5 Investigating the variation in magnetic flux density with angle, using the magnetic field from a permanent magnet with a Hall probe

> **Investigations**
>
> - Investigate the strength and direction of magnetic fields, such as those of magnadur magnets on a yoke, solenoids and transformers, using a Hall probe.
> - Apply a magnetic field to an ionic solution (for example, $CuSO_4(aq)$) and look for separation of ions.

Using a Geiger–Müller (GM) tube

A Geiger–Müller (GM) tube is used to detect ionizing radiation. It consists of a cylindrical container filled with an inert gas (for example, argon) at low pressure. There is a wire stretched along the axis of the cylinder. A high voltage is applied between the wire and the cylinder. When alpha or beta radiation enters the tube through the GM tube 'window', the particles collide with and ionize an atom of the gas in the tube. Because of the high voltage between the wire and the cylinder, electrons and positive ions ($Ar^+(g)$) accelerate towards their oppositely charged electrodes, colliding with and ionizing other atoms in their path. The resulting cascade of ions produces an electrical pulse in the detector.

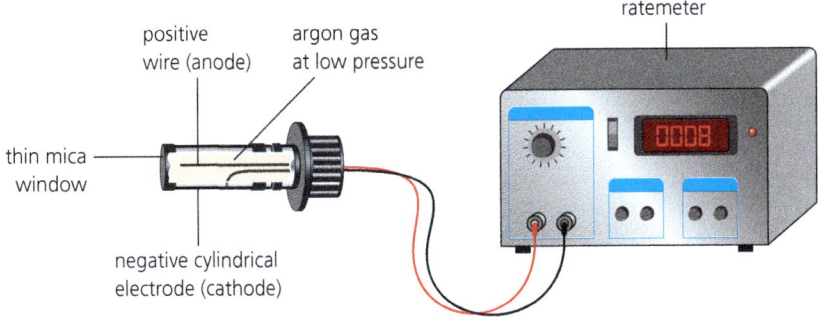

Figure 3.6 A Geiger–Müller tube and ratemeter

Activity (decays per second, in becquerel, Bq) of a radioactive source can be measured using a GM tube connected to a ratemeter (Figure 3.6). The GM tube detects each time a particle, emitted in an individual nucleus decay event, enters the tube. The detected count rate is displayed on the ratemeter as the number of counts per second (or per minute).

A GM counter cannot differentiate which type of radiation is being detected and cannot be used to determine its energy.

Expert tip
When taking count readings, the longer the count, the lower the uncertainty. It can be shown that the uncertainty in a total count of N is $\pm\sqrt{N}$. A total count of 400 will have an uncertainty of ±20, or ±5%.

Examiner guidance
Background radiation is always present. This background count rate should always be subtracted from a measured count rate to obtain the rate from the source alone. Obtain a background count rate over a 5-minute sample time.

Investigations
- Verify the inverse square law for the effect of distance from the source on the count rate.
- Determine the relationship between the half-value thickness and the density of different absorber materials.
- Determine the half-life of radioactive substances, for example, protactinium, by plotting a half-life decay curve.

You must consult your physics teacher before working with radioactive isotopes.

RESOURCES
- http://practicalphysics.org/managing-radioactive-materials-schools.html
- https://www.tf.uni-kiel.de/matwis/amat/iss/kap_2/articles/beer_article.pdf
- http://practicalphysics.org/gamma-radiation-inverse-square-law.html

Examiner guidance
Different countries have different national laws (that follow internationally agreed principles of radiological protection) to control how radioactive materials are acquired, used and disposed of. Many countries ban school experiments with radioactive sources, in which case simulations may be used. For example, it is possible to use rice grains as improvised materials for demonstrating radioactivity. (See *School Science Review*, September 1992, 74(266), 106–109.)

Using a current balance

A U-shaped permanent magnet is placed on a top pan balance (Figure 3.7). An insulated wire is clamped so that it runs perpendicularly to the field between the poles. It is part of a series circuit with a dc supply, a rheostat, an ammeter and a simple switch. When the switch is turned on, the balance reading is observed to change because the current creates a magnetic field that causes a repulsive force between the magnet and the wire. By Newton's third law of motion, an equal force is exerted on the magnet, which is detected by the change in the mass reading, Δm.

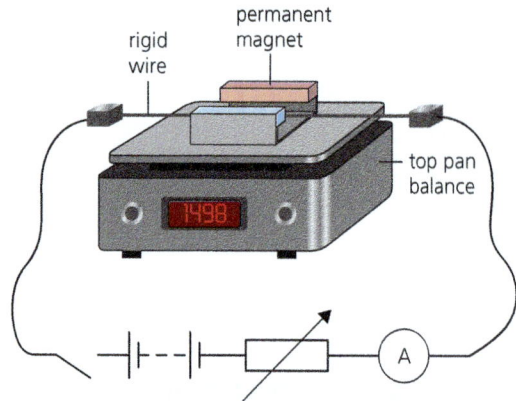

Figure 3.7 Principle of the current balance

This magnetic force F is equal to $\Delta m \times g$. The variation of F with the current I may be determined from the relationship $F = B \times I \times L$, where the current, I, is perpendicular to the field of flux density, B and L is the length of the wire in the magnetic field between the two poles of the magnet. The direction of the force, as predicted by Fleming's left hand rule, may be verified by observing whether the mass reading increases or decreases for a specific direction of current flow.

Worked example

A 45.0 mm length of wire in which a current of 1.00 A flows is at right angles to a magnetic field of flux density 0.12 T. Calculate the force acting on the wire in mN.

$F = BIL = 0.12 \text{ T} \times 1.00 \text{ A} \times 4.5 \times 10^{-2} \text{ m} = 5.4 \times 10^{-3} \text{ N} = 5.4 \text{ mN}$

Investigations

For a parallel-wire current balance (two parallel identical wires carrying electric current):

- plot force against current and also force against separation
- investigate variation in lengths of parallel wire against force.

Using a ripple tank

A ripple tank (Figure 3.8) can demonstrate properties of waves, including reflection, refraction, diffraction and interference. The tank is shallow with sloping sides (to reduce reflection when the waves hit the sides of the tank) and with a transparent bottom so that when a light source is mounted above the tank, a magnified image of the water waves is projected onto a screen below the tank.

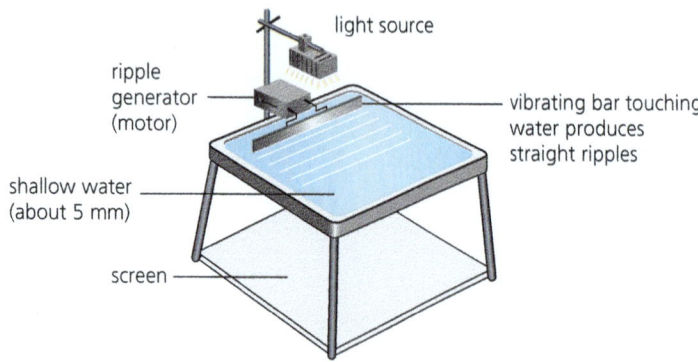

Figure 3.8 A ripple tank

The movement of the waves can be 'stopped' for certain observations and measurements using a stroboscope – either a simple hand wheel with regularly spaced slits or an electronic stroboscope. The stroboscope gives a series of 'snapshot' views of the wave, timed at such a frequency that each snapshot is exactly one period after the previous one. This means that you see each wave crest in exactly the same position as the preceding wave crest, so the wave appears to be standing still.

> **Examiner guidance**
>
> In all work with flashing lights, you must be aware of any student working in the lab who suffers from photo-induced epilepsy. Your physics teacher must ask any known epileptic whether an attack has ever happened with flashing lights. If so, the student should be asked to leave the lab.

Continuous ripples (waves) are generated using an electric motor and a bar. The bar gives straight ripples if it just touches the water or circular ripples if it is raised and has a small ball fitted to it which touches the water surface at one point. A range of accessories may then be placed in the path of the waves to show properties of waves.

> **Investigations**
>
> Investigate wave properties including phase and velocity for different water depths and in the presence of obstacles of various shapes, for example, plane mirrors and concave refractors/reflectors (also the effect of focal length).

> **RESOURCES**
>
> https://www.rapidonline.com/eisco-ph0767a-ripple-tank-kit-tank-size-400-x-400mm-52-3383

> **Worked example**
>
> A water wave in a ripple tank travels from a shallow to a deep region. The wavelength and speed in the shallow region are 2.50 cm and 5.00 cm s^{-1}, respectively. If the wavelength in the deep region is 6.00 cm, determine the relative index of refraction from shallow to deep water and the speed of the wave in the deep water.
>
> Let 's' denote shallow and 'd' denote deep.
>
> $$_s n_d = \frac{\lambda_s}{\lambda_d} = \frac{2.50 \text{ cm}}{6.00 \text{ cm}} = 0.4167 = 0.42$$
>
> $$c_d = \frac{c_s}{_s n_d} = \frac{5.00 \text{ cm s}^{-1}}{0.4167 \text{ cm}} = 12 \text{ cm s}^{-1}$$

Using a sonometer

A sonometer (Figure 3.9) consists of a hollow wooden box about 1 metre long, the 'sound board'. A thin metal wire is stretched over it; one end of the wire is fastened at the edge of the sound board and the other end passes over a pulley and carries a mass hanger. The weight of these masses produces tension in the wire and presses it against the two bridges. One of these bridges is fixed and the other is movable.

The wire can be made to vibrate by plucking it, and the length of the vibrating wire can be changed by moving the bridges. The wall of the sound board contains holes so that the air inside the sound board remains in contact with the air outside. When the wire vibrates, these vibrations reach (via bridges) the upper surface of the sound board and the air inside it. The air outside the sound board also begins to vibrate and a loud sound is heard. The transverse waves on the wire will transmit longitudinal air waves.

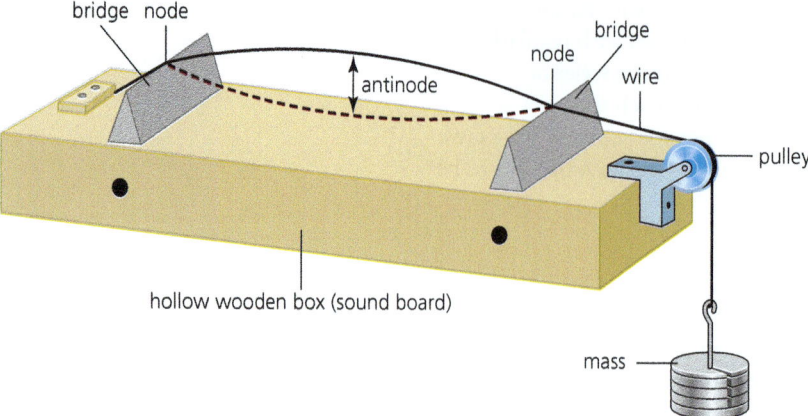

Figure 3.9 A sonometer

> **Expert tip**
>
> Place a small inverted paper 'V' centrally on the wire and place a tuning fork on the fixed bridge or the sound board. Adjust the wire length until resonance occurs and the paper rider is thrown off. The frequency is now known.

When the sonometer wire is plucked in the middle, it vibrates in its fundamental mode with a natural frequency $f_0 = \frac{1}{2l}\sqrt{\frac{T}{m}}$, where l is the length of the wire between the bridges, T is the tension in the wire and m is the mass per unit length of the wire.

> **Worked example**
>
> A sample of wire was being tested for use as a guitar string. It had a mass per unit length of 0.004 kg m^{-1}. A 64.0 cm length of the wire was fixed at both ends and plucked. The fundamental frequency of the sound produced was found to be 173 Hz. Calculate the tension in the wire.
>
> $4l^2 f^2 = \frac{T}{m}$
>
> $T = m \times 4l^2 f^2$
>
> $T = (0.004) \times (4) \times (0.64)^2 \times (173)^2 = 196$ N

> **RESOURCES**
>
> Investigations using a sonometer can be found at these websites.
>
> To determine the frequency of ac using a sonometer:
> - http://amrita.olabs.edu.in/?sub=1&brch=6&sim=151&cnt=1
>
> Investigating transverse waves in a stretched string using a sonometer:
> - https://www.youtube.com/watch?v=GTnPEtksTEc

If r is the radius of the wire, ρ the density of the material of the wire and M the mass suspended from the wire, then $m = \pi r^2 \rho$ and $T = Mg$, hence

$$f_0 = \frac{1}{2l}\sqrt{\frac{Mg}{\pi r^2 \rho}}$$

> **Investigations**
>
> - Establish the relationship between the frequency of the sound and the length, tension and mass per unit length of the wire.
> - Determine the frequency of ac (from a signal generator) using a sonometer with a horseshoe magnet and piece of paper on the (non-magnetic) wire.

Using a ray box and optic bench

A ray box is an instrument that produces single or multiple rays of visible light, which can be made visible on paper. The bulb and lens can be moved relative to each other so that the emergent rays may be made convergent, divergent, or parallel if the bulb is at the focus of the lens.

Experiments with a ray box or, alternatively, a lamp and slit (Figure 3.10), are most effective in a darkened laboratory. The reflected and refracted light rays then show up well on white paper.

An accurate optical bench (Figure 3.11) consists of a long, rigid ruler with a linear scale applied to it. Holders for light sources, lenses and screens are placed on the apparatus so that image formation can be observed.

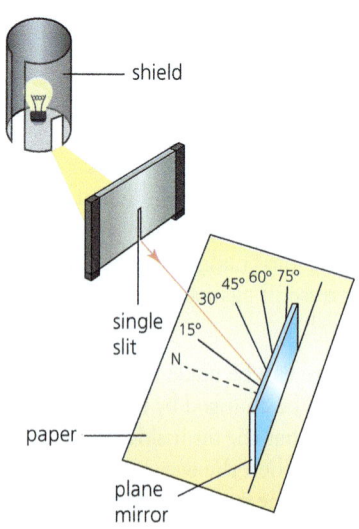

Figure 3.10 Investigation of reflection using a lamp, slit and plane mirror

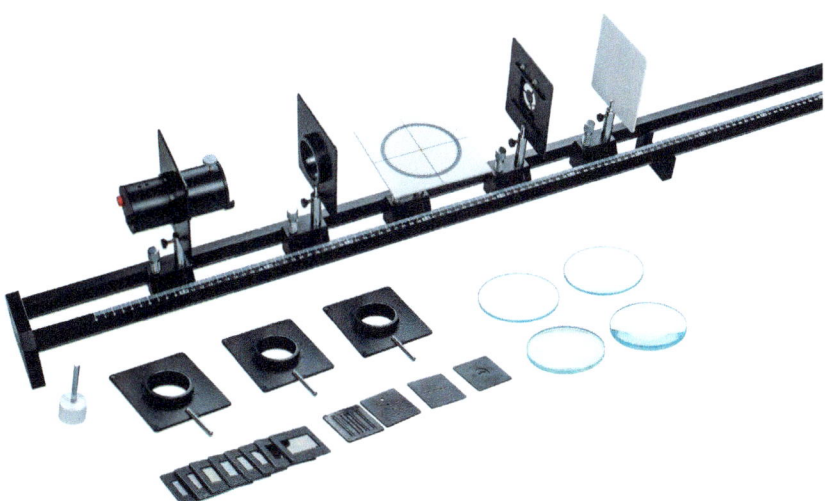

Figure 3.11 An optical bench with lenses (photo courtesy of Philip Harris)

> **Expert tip**
>
> Experiments are easier to arrange if cylindrical mirrors and lenses are used instead of spherical ones. Since the light rays are confined to one plane this does not invalidate the experiment.

> **Expert tip**
>
> The spherometer is an instrument used to measure the radius of curvature of a spherical or cylindrical surface and the thickness of a very thin glass plate (see https://www.a3bs.com/product-manual/U15030_EN.pdf on the use of a precision spherometer).

> **Investigations**
>
> Investigate the properties of thin and thick lenses and how the lens power $\left(\frac{1}{f}\right)$ is related to the curvature of its faces and the refractive index of its material.

> **Examiner guidance**
>
> Rules for ray diagrams involving curved mirrors and lenses:
> - A ray of light that is parallel to the principal axis will be reflected (or refracted) through the principal focus.
> - A ray of light that passes in through the principal focus will be reflected (or refracted) parallel to the principal axis.
> - A ray of light that is incident on a mirror along a radius of curvature will be reflected back along its own path.
> - A ray of light that strikes the centre of a lens will pass straight through.

Using a spectrometer

A spectrometer is an instrument for producing spectra and for measuring the deviation of refracted or diffracted light rays and hence allowing the wavelength of light to be calculated. The spectrometer consists of three main parts: collimator, table and travelling telescope (Figure 3.12).

- The collimator produces a beam of parallel light from the source. It is a tube with a lens at one end and an adjustable slit at the other.
- The table is a rotating platform for the diffraction grating or prism. It includes three levelling screws to ensure that the diffracted image is in the centre of the field and a vernier scale for very accurate measurement of position.
- The travelling telescope receives the diffracted or refracted beam of light and, via the cross wires in the eyepiece, the angle of diffraction or refraction can be determined.

> **Investigations**
>
> Investigate the effect of the angular spread of the visible spectrum (1st and 2nd orders) for prisms and diffraction gratings.

■ Adjusting the spectrometer

The eyepiece of the telescope is adjusted until the cross wires can be seen clearly. The telescope is then adjusted to receive parallel light rays, by focusing on vertical lines on some distant object (such as a brick wall) using the focusing screw. There should be no parallax between the distant lines and the cross wires.

The collimator is adjusted by putting it in line with the telescope and viewing a bright light source, for example, a sodium lamp. The collimator slit position is adjusted until a sharp image is seen through the telescope, which means the collimator is producing parallel light rays.

The table is adjusted to make sure that the distant lines are vertical to the axis of the collimator and its face is perpendicular to the axis of the collimator. This is done with the leveling screws as follows.

Place the grating on the table as shown in Figure 3.12.

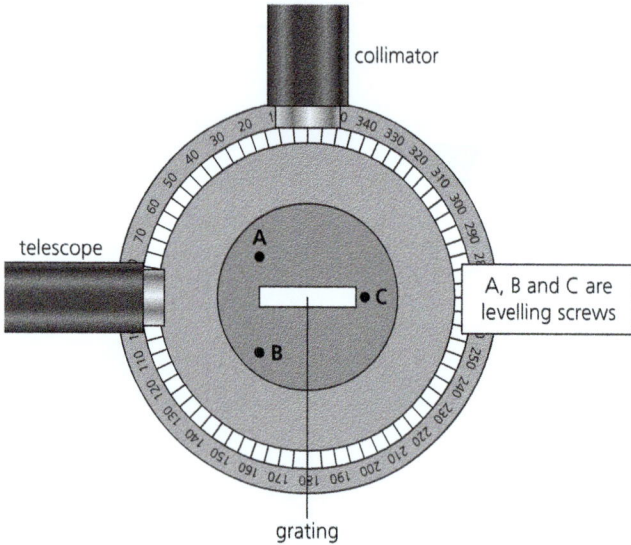

Figure 3.12 Plan view of spectrometer table

Rotate the telescope until it is at 90° to the collimator. Rotate the table until an image is seen in the eyepiece. Adjust screw A until the image occupies the same position in the field as it did when the telescope and collimator were in line.

Rotate the grating through 45°. Rotate the telescope back past the original position until an image is seen in the eyepiece. Adjust screw C until the image occupies the same position in the field as previously.

Using lasers

Lasers provide an opportunity to demonstrate a wide range of optical principles and phenomena in the study of light. You may choose to use a laser in your individual investigation.

Before using a laser you must get the approval of your physics teacher and you must carry out a risk assessment.

■ Safe use of lasers

Laser light has special properties; for example, extreme brightness, low divergence and significant phase coherence. It needs to be used with extreme care.

- Never look directly into the laser beam or into any direct reflections of the beam, and never aim it into the eyes of any other person in the lab. When using a laser beam in a horizontal plane, keep it away from eye level. The presence of the laser beam should always be detected by a white card and never with the naked eye.

- Use as high a level of background lighting as is possible for the experiment being performed. This reduces the size of your eye's pupil to a minimum and provides some protection.
- Avoid reflections of the beam from the surface of shiny objects if the laser is being moved around. Cover or remove any bright reflective objects or block reflected light with non-reflective objects such as black paper.
- Be careful when placing lenses, mirrors or other beam deviators in the path of the beam. Move slowly, and try to predict the direction of the beam after it 'interacts' with the lens or mirror.

Investigations

- Study the intensity profile across the laser beam, and the intensity variation with distance.
- Investigate polarization of the beam.
- Use optical interference to determine the wavelength of red and green laser light. Use the wavelength to determine the track spacing on a standard CD.

Using an air track

An air track (Figure 3.13) produces almost frictionless linear motion over a distance of about 1.5 metres. The motion of light aluminium gliders on a very thin cushion of air can be studied.

The glider track is a tube of triangular cross section with many tiny holes; air is pumped into the tube at one end and leaves through the holes in the two upper surfaces of the track. The air does not cause the motion of the gliders but supports them just off the track surface. The gliders are therefore free to move on the track, to collide with each other and with elastic bumpers at the ends of the track, or to be pulled and pushed by string and pulley systems, with very little friction.

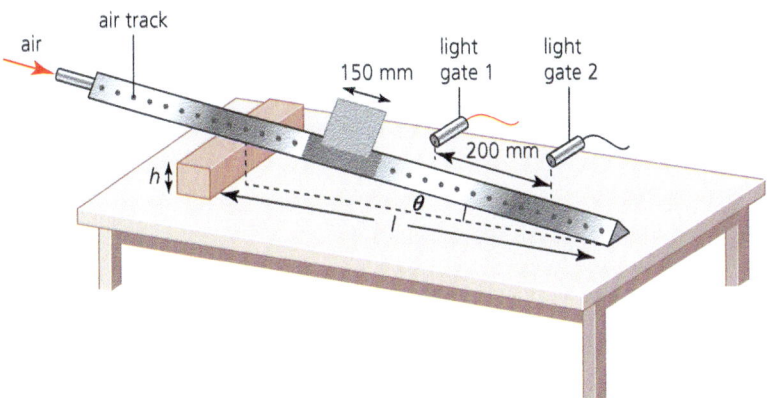

Figure 3.13 An arrangement for investigating the acceleration of a glider on a tilted air track

Expert tip

The air track and gliders operate best if they are clean and their surfaces are smooth with no bumps or nicks.

RESOURCES

You can find investigations using an air track described at these websites.

- http://www.iecpl.com.au/z_exp/mf0105z-001exp.pdf
- https://www.unr.edu/Documents/science/physics/labs/180/05_Force_and_Acceleration_on_the_Air_Track.pdf
- http://userhome.brooklyn.cuny.edu/kshum/documents/5.AccAirTrack.pdf
- http://users.df.uba.ar/sgil/physics_paper_doc/papers_phys/mechan/airdrag1.pdf

Worked example

An air-track glider is released from rest and the passage of its interrupter card, of length 150 mm, through two light gates is timed. Passage times of $t_1 = 1.48$ s and $t_2 = 0.25$ s are recorded. The distance between the light gates is 200 mm (Figure 3.13). Calculate the average velocity of the glider as it passes through each light gate and its acceleration.

$$v_1 = \frac{0.150 \text{ m}}{1.48 \text{ s}} = 0.101 \text{ m s}^{-1}$$

$$v_2 = \frac{0.150 \text{ m}}{0.25 \text{ s}} = 0.600 \text{ m s}^{-1}$$

Using $v^2 = u^2 + 2as$:

$$a = \frac{(0.600 \text{ m s}^{-1})^2 - (0.101 \text{ m s}^{-1})^2}{2 \times 0.200 \text{ m}} = 0.87 \text{ m s}^{-2}$$

> **Investigations**
>
> - Investigate momentum during collisions.
> - Collect data to determine the relationship between the net force exerted on an object, its inertial mass and its acceleration.

Using a function generator

A signal or function generator is a device containing an RC oscillator, which can produce various patterns of voltage at a variety of frequencies and amplitudes. A common use is to test the response of circuits to a known input signal. Most function generators allow you to generate sine, square or triangular ac function signals. You can view the signals produced by connecting the signal generator to an oscilloscope.

> **Expert tip**
>
> It is easy to misread scales on an analogue device which has a 0–10 scale above the graduations, and a 0–3 scale underneath.
>
> Take care when using a signal generator, for example, where an analogue 1–10 dial is used in conjunction with multiplier knobs such as 1–10 kHz.
>
> With an instrument such as an oscilloscope you have to be careful to check whether a multiplier knob such as ×10 has been selected.

> **RESOURCES**
>
> https://www.iosrjen.org/Papers/vol2_issue5/G025971978.pdf

4 Mandatory practicals

Determination of the acceleration of free-fall due to gravity

■ Essential theory

A free-falling object falls vertically under the force of gravity, with no other forces acting. However, an object falling through air will experience fluid resistance or air resistance (a frictional force). If the object is small and a high density and its speed is not excessive, it can be regarded as being in free-fall. The equations of uniform motion can be applied to the accelerating object, using $a = g$, the acceleration of free-fall.

■ Suggested methods

A steel sphere is released from an electromagnet and falls under gravity. As it falls downwards, it passes through light gates which switch an electronic timer on and off (Figure 4.1). The acceleration of free-fall can be determined from the values of the time intervals for different distances between the gates. This is an experiment that requires timing to be accurate to the nearest one-hundredth of a second.

The analysis of the results relies on the application of the following equation of motion (for uniform acceleration): $s = ut + \frac{1}{2}at^2$, where s is the distance, u is the initial velocity, t is the time and a is the acceleration.

Another method of determining g uses an electromagnet and a trap door. The distance between the electromagnet and the trap door is measured. An electronic timer starts when the ball is released and stops when the ball hits the trap door.

Since u is zero and the cause of the acceleration is the Earth's gravitational field then the equation of motion can be rewritten as $s = \frac{1}{2}gt^2$, which can be rearranged to give $2s = gt^2$. This is now in the form $y = mx + c$ and hence plotting $2s$ on the y-axis and t^2 on the x-axis will result in a straight line of gradient g.

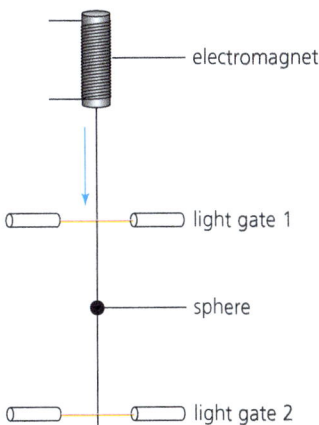

Figure 4.1 Determination of the acceleration of free-fall

> **Examiner guidance**
>
> When using light gates the passage time of the moving object may be very short if the object is travelling fast. Assuming the electronic timer measures to 0.01 s, the time of passage should be at least 0.2 s to ensure the experimental error is less than 5%. Therefore the object should not be allowed to travel too fast; air resistance is also more significant for greater speeds.

ACTIVITY

1 Table 4.1 shows a set of experimental data from a practical to determine g by a free-fall method using the trap-door approach. The times here are the average of three values for each height of fall.

Height of free fall, h/m ±0.005 m	Average time of free fall, t/s ±0.01 s	t^2/s^2
0.600	0.39	
0.800	0.45	
1.000	0.48	
1.200	0.53	
1.400	0.57	
1.600	0.60	

Table 4.1 Experimental data from free-fall investigation

Complete the table to show values of t^2 and plot a graph of $2s$ on the y-axis against t^2 on the x-axis. Derive a relationship for the gradient of the graph and determine a value for g. Suggest why the graph does not begin at the origin.

Examiner guidance

In practice, all raw data must be included in your IB Physics internal assessment report.

RESOURCES

Measuring acceleration of free fall:
- https://nustem.uk/activity/measuring-g/
- http://cfinndcs.wdfiles.com/local--files/mechanics/Exp-Measurement%20of%20Acceleration%20Due%20to%20Gravity%20-%20Free-fall%20Method.pdf
- https://www.youtube.com/watch?v=1XnHQJwEI4U
- http://practicalphysics.org/measurement-g-using-electronic-timer.html

Examiner guidance

If you are raising an uncertain number to a power n (for example, squaring it or taking the square root), then the resulting number has a percentage uncertainty n times the percentage uncertainty in the original number. Thus if $t = 2.36$ s ± 0.04 s, then the percentage uncertainty in t^2 is ±(2 × 1.695%) = ±3.39%.

Determining specific heat capacities

Essential theory

When a substance is heated, a temperature change is generally observed. Different substances require different amounts of thermal energy to cause the same temperature rise in the same mass. The specific heat capacity of a substance is the thermal energy required to produce a 1 °C rise in 1 kg of the substance. The unit is therefore J kg^{-1} °C^{-1}.

The thermal energy, Q, that is added or removed to cause a temperature change, ΔT, is $Q = mc \Delta T$, where m is the mass of the substance and c is its specific heat capacity, assumed to be constant over the temperature change, ΔT.

The method of mixtures

To determine the specific heat capacity of a substance, the method of mixtures is often used (Figure 4.2). A container ('calorimeter') of known specific heat capacity, c_c, and mass, m_c, is partially filled with a mass, m_w, of water at a temperature, T_1.

A bolt of mass, M, of a substance of unknown specific heat capacity, c, is heated to a higher temperature, T_b, and then quickly transferred to the water in the calorimeter. The temperature of the calorimeter and the water contained in it quickly rises to a value, T_2. It then slowly begins to fall as heat is lost to the surrounding air. (If an insulated glass beaker is used as the calorimeter it can be assumed that no thermal energy is transferred from the water during the rise in temperature.)

The hot substance has therefore transferred $Mc(T_b - T_2)$ joules of thermal energy to the calorimeter and the water inside. If no heat losses occur, this must be equal to the thermal energy gained by the water and calorimeter, which is $(m_c c_c + m_w c_w)(T_2 - T_1)$ and hence

$$Mc(T_b - T_2) = (m_c c_c + m_w c_w)(T_2 - T_1)$$

and the specific heat capacity c of the bolt can be determined.

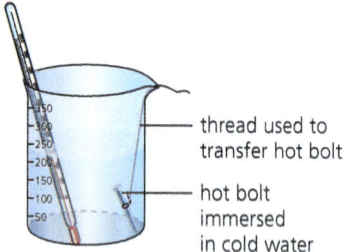

Figure 4.2 The method of mixtures used to determine the specific heat capacity of a steel bolt

Electrical methods

A simple way of determining the specific heat capacity of a substance is to supply a known amount of thermal energy from an electric heater placed inside the substance. For solids, a hole needs to be drilled in the substance to allow the immersion heater to be fitted inside, ensuring good thermal contact.

In the two experiments shown, one involving a metal (Figure 4.3) and the other water (Figure 4.4), a joulemeter is used to measure the energy transferred directly. The energy transferred, Q, could also be calculated from the expression $Q = Pt$ or $Q = VIt$. If the mass of the specimen (solid or liquid) is m and its specific heat capacity c, then $Q = mc\,\Delta T$; rearranging, $c = \frac{Q}{m\,\Delta T}$.

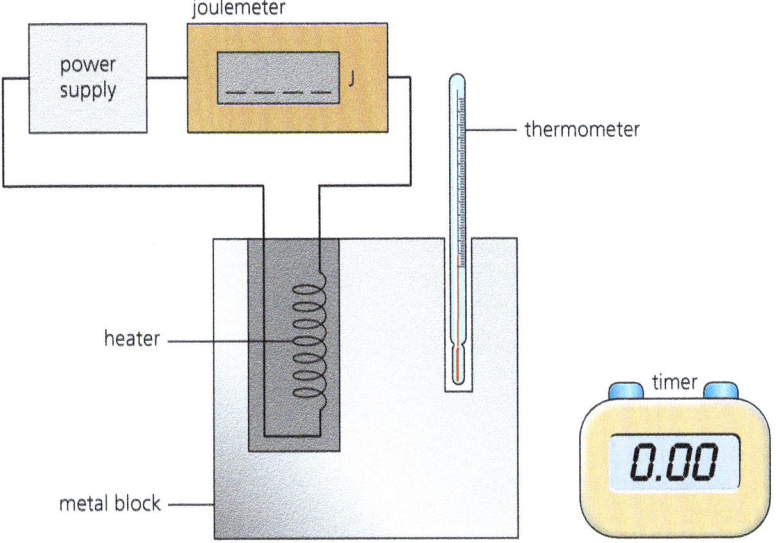

Figure 4.3 Determining the specific heat capacity of a metal by an electrical method

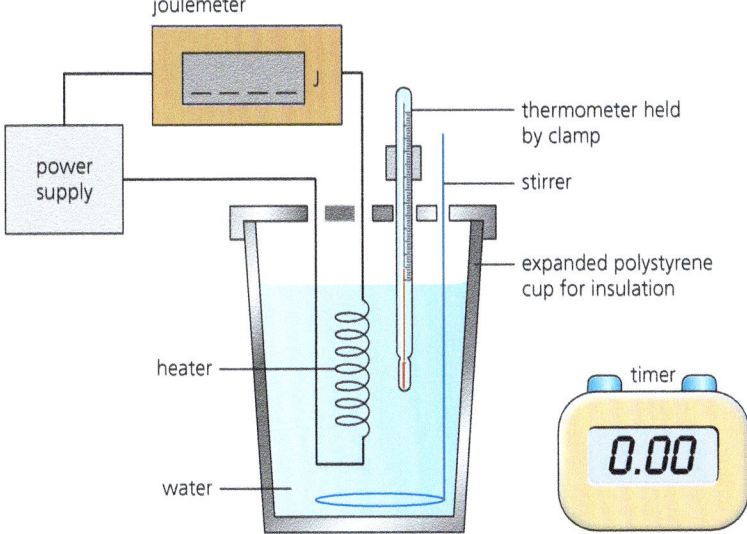

Figure 4.4 Determining the specific heat capacity of water by an electrical method

Examiner guidance

This simple method can be used for liquids or solids, although in the case of a liquid, allowance has to be made for the heat capacity of the container. Also, the liquid should be stirred to evenly distribute the heat energy throughout its volume. This is necessary since liquids are poor thermal conductors.

Expert tip

- In electrical methods, where a constant current is required, always include a rheostat in the circuit.
- When a heating coil is used it must always be completely covered with liquid to prevent it 'burning out'.

> **Worked example**
>
> A value for specific heat capacity of a metal block can be determined from the experimental results given below. State and explain which of the five measurements produces the greatest uncertainty in the calculated value of the specific heat capacity.
>
> | | |
> |---|---|
> | Power | 2000 W ± 10 W |
> | Time of heating | 300 s ± 1 s |
> | Mass of metal block | 5.0 kg ± 0.2 kg |
> | Final temperature of metal block | 50.0 °C ± 0.5 °C |
> | Change in temperature | 30 °C ± 1 °C |
>
> The calculated percentage uncertainties are, respectively, 0.5%, 0.3%, 4%, 1% and 3.3%, hence the mass measurement produces the greatest uncertainty in the calculated value of the specific heat capacity.

Determining specific latent heats

■ Essential theory

When a substance is heated and it changes state, the temperature remains constant while the change of state occurs. Two common changes of state are from solid to liquid (melting or fusion) and from liquid to gas (boiling or vaporization). The amount of thermal energy required to cause all of the substance to change state is the latent heat. For a solid to liquid change (at constant temperature), this is referred to as the latent heat of fusion; for a liquid to gas state change, it is the latent heat of vaporization.

The specific latent heat of fusion or vaporization, L, is the amount of thermal energy required for 1 kg of the substance to change state.

$Q = mL_f$ or mL_v, where L_f or L_v is the specific latent heat of fusion or vaporization.

■ Determining specific latent heat of fusion for ice

The specific latent heat of fusion for ice can be experimentally determined through the measurement of the mass of water, m, produced when energy, Q, is transferred to melting ice.

An electric immersion heater is inserted into a funnel and crushed ice is packed around it (Figure 4.5). To correct for thermal energy transferred from the surroundings, first, with the heater off, the melted ice is collected in a beaker for time, t (for example, 5 minutes); the beaker and the melted ice is weighed and its mass, m_1, determined. The beaker is then emptied, the heater switched on, and melted ice collected for the same time period, t. The beaker with the melted ice is reweighed and its mass, m_2, determined. The mass of ice melted by the heater is given by $m_2 - m_1$.

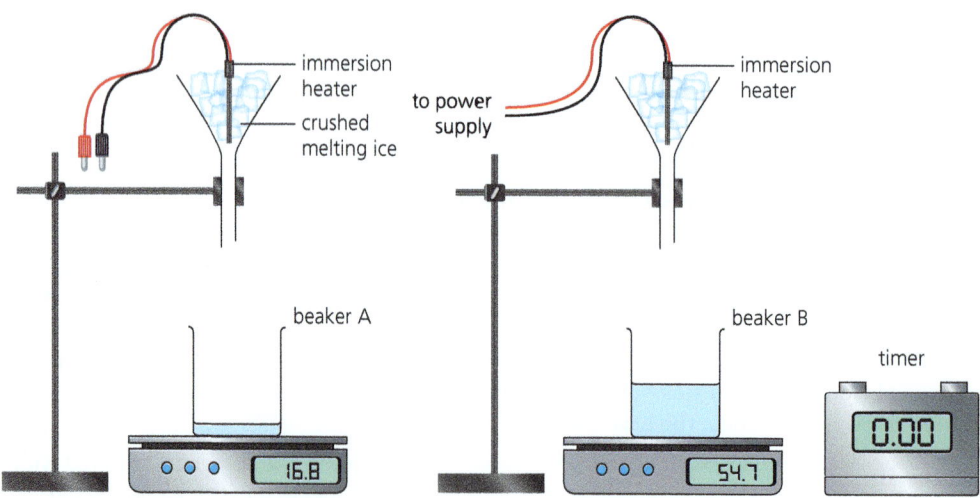

Figure 4.5 Apparatus for determining the specific latent heat of fusion for ice

The electrical energy supplied by the heater of power, P, is $Q = P \times t$. Alternatively, a joulemeter can be used to record Q directly. The specific latent heat of fusion for ice, L_f, can be calculated from: $Q = (m_2 - m_1) \times L_f$.

■ ACTIVITY

2 In an experiment to measure the specific latent heat of fusion of ice, warm water was placed in an aluminium calorimeter. Crushed, dried ice was added to the water and melted. The following results were obtained; for simplicity the uncertainties have been omitted.

> Aluminium = 900 J kg^{-1} °C^{-1}
> Water = 4200 J kg^{-1} °C^{-1}
> Mass of aluminium calorimeter = 77.20 g
> Mass of water = 92.50 g
> Initial temperature of water = 29.4 °C
> Assumed temperature of ice = 0.0 °C
> Mass of ice = 19.2 g
> Final temperature of water = 13.2 °C
> Room temperature = 21.0 °C

a Suggest the advantage of having the room temperature approximately halfway between the initial temperature of the water and the final temperature of the water.

b Outline how the mass of the ice was found.

c Calculate a value for the specific latent heat of fusion of ice.

d The accepted value for the specific latent heat of fusion of ice is 3.3×10^5 J kg^{-1}. Suggest two reasons why the experimental measurement differs from the literature value and calculate the percentage error.

> **Key definition**
>
> **Literature value** – An accepted value from the physics literature of a physical constant or experimental measurement.

■ Determining specific latent heat of vaporization for steam

The specific latent heat of vaporization for steam can be calculated by measuring the mass of vapour (steam), m, produced when thermal energy, Q, is transferred to boiling water.

Water in a flask (Figure 4.6) is electrically heated to boiling point. Steam passes out through the holes in the top of the flask, down the outside of the flask and into the inner tube of a condenser, where it changes back to liquid, and is collected in a conical flask.

After the water has been boiling for a period of time, it becomes enclosed by a 'jacket' of vapour at the boiling point, which helps to reduce loss of heat to the surroundings. The rate of vaporization becomes equal to the rate of condensation, and the electrical energy is only being used to transfer latent heat to the water (not to raise its temperature).

The electrical energy, Q, supplied by the heater of power, P, is given by $Q = P \times t = ItV$, where I is the steady current through the heater and V is the pd across it. Alternatively, a joulemeter can be used to record Q directly. If a mass of water, m, is collected in time, t, then the specific latent heat of vaporization, L_v, can be calculated from $Q = m \times L_v$.

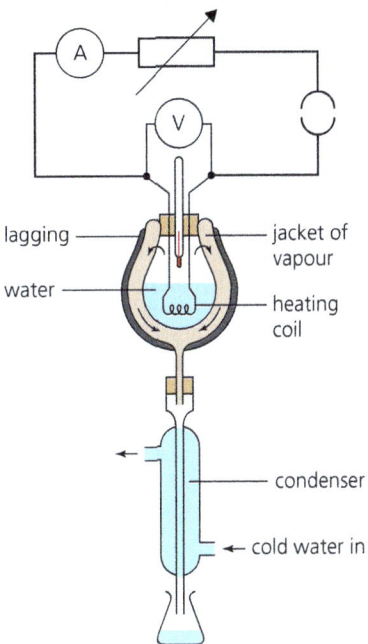

Figure 4.6 Apparatus to determine the specific latent heat of vaporization for steam

Worked example

Two experimental runs were carried out using the apparatus in Figure 4.6. The results are shown in Table 4.2.

	Power of heater/W	Time/min	Mass of beaker and water at the start of the experiment/g	Mass of beaker and water at the end of the experiment/g
Experiment 1	36	10	178.8	169.4
Experiment 2	50	10	169.4	156.3

Table 4.2 Data from vaporizing experiments

Experiment 1:

electrical energy input $- Q_s = \Delta m L_v$, where Q_s is the thermal energy lost to the surroundings:

$36 \text{ W} \times 600 \text{ s} - Q_s = 9.4 \text{ g} \times L_v$

$21\,600 \text{ J} - Q_s = 9.4 L_v$

Experiment 2:

Using the same equation:

$50 \text{ W} \times 600 \text{ s} - Q_s = 13.1 \text{ g} \times L_v$

$30\,000 \text{ J} - Q_s = 13.1 L_v$

Subtracting the first equation from the second:

$30\,000 \text{ J} - 21\,600 \text{ J} = (13.1 - 9.4) L_v$

$L_v = 2.27 \times 10^3 \text{ J g}^{-1}$

Expert tip

- Most of the errors in these experiments arise from inaccurate temperature readings, so it is advisable to use a ±0.1 °C thermometer or temperature probe.
- Avoid small temperature changes in the experiments because these have large percentage errors.
- Record temperature readings at eye level to avoid parallax error. A small magnifying glass can help in accurately determining the thermometer reading.
- A stirrer, if used, should be of the same material as the calorimeter. The mass of the stirrer should be included in the mass of the calorimeter. Always stir the liquid in a calorimeter before taking a reading, and take the highest or lowest (whichever it may be) steady temperature.
- Substances (for example, crushed ice) added to water in a calorimeter should be transferred rapidly, but without splashing.

Examiner guidance

For your internal assessment:

- times should be reported in seconds, not minutes
- uncertainties should be recorded and error propagation performed (see Chapter 5)
- the experimental value for the specific latent heat of vaporization should be compared to a referenced literature value, and absolute and percentage differences calculated.

■ Dealing with heat losses in heat capacity and latent heat experiments

The approach in the worked example above shows that where the heat loss is assumed to be the same in two runs, it can be eliminated. Avoid having a large temperature difference between the calorimeter and its surroundings, because the rate at which heat flows from one place to another (for example, calorimeter and surroundings) depends on the temperature difference between the two places.

The use of a highly polished calorimeter reduces heat loss by radiation. Having a lid on the calorimeter reduces heat losses by convection and evaporation.

A polystyrene cup is a useful object in heat experiments because it has a negligible mass and heat capacity, and almost all the thermal energy goes into the liquid contained in the cup.

Insulation or lagging (draught proofing) is all-important in experimental work on heat, to reduce thermal energy lost to (or gained from) the surroundings. A piece of 'aero board' (expanded polystyrene) placed underneath the calorimeter as a stand will help.

To minimize the effect of inevitable heat losses/gains, start with the calorimeter a few degrees below room temperature and finish with it an equal value above room temperature (or vice versa if the experiment involves cooling).

Gas laws

■ Essential theory

Three important bulk properties of gases are pressure, volume and temperature; they are independent of the nature of the particles under normal conditions, described as 'ideal'. The gas laws describe the relationships between these variables.

- Boyle's law gives the relationship between pressure and volume at constant temperature:

  the pressure of a fixed mass of gas is inversely proportional to its volume if the temperature remains constant.

- The pressure law gives the relationship between pressure and temperature at constant volume:

  for a fixed mass of gas (at constant volume) its pressure is directly proportional to its absolute temperature.

- Charles' law gives the relationship between volume and temperature at constant pressure: for a fixed mass of gas (at constant pressure) its volume is directly proportional to the absolute temperature.

Ideal behaviour is assumed.

■ Boyle's law

Boyle's law apparatus (Figure 4.7a) is used to study the effect of pressure changes on a fixed volume of gas. The gas being tested is compressed (at constant temperature) using the pump. Its pressure is recorded from the pressure gauge because it is equal to the pressure of the oil in the reservoir. The volume of the gas is recorded from the scale. The pressure is increased in several stages, so that a number of pairs of pressure/volume readings are recorded.

A data-logging approach with a pressure sensor can also be used (Figure 4.7b).

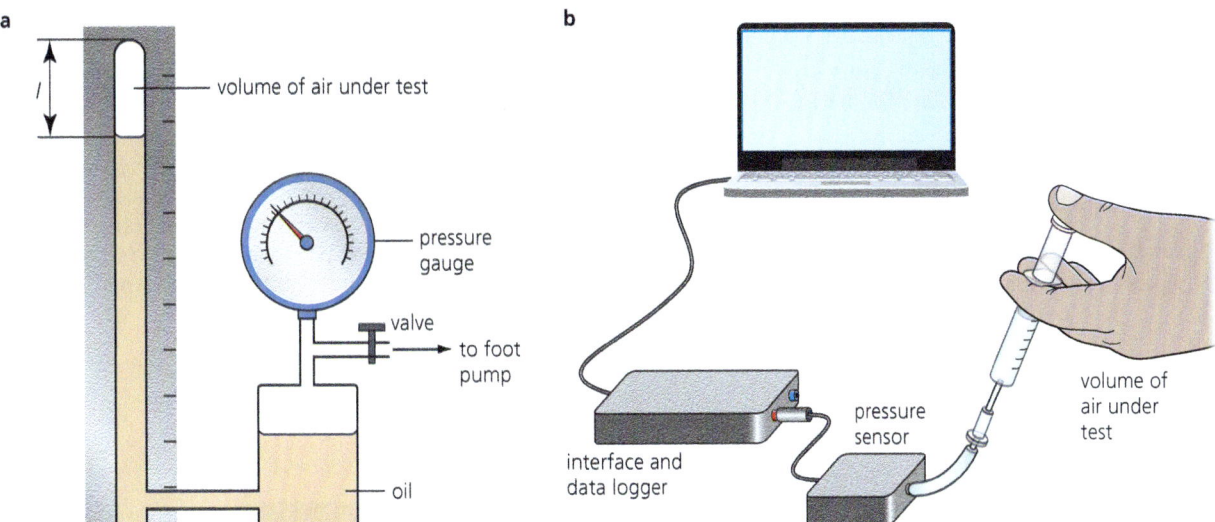

Figure 4.7 Investigating Boyle's law

If you plot a graph of pressure against volume using the results, you obtain a hyperbolic curve (Figure 4.8). The graph shows that if the pressure is doubled, the volume is halved. That is, pressure is inversely proportional to volume. In symbols, $p \propto \frac{1}{V}$ or $p = \text{constant} \times V^{-1}$ and therefore $pV = \text{constant}, k$. If several pairs of readings are recorded it can be checked that $p_1V_1 = p_2V_2 = \text{constant}$. To verify that $p = \frac{k}{V}$ you can plot a graph of p against $\frac{1}{V}$. If this shows a proportional relationship then the correlation is proven.

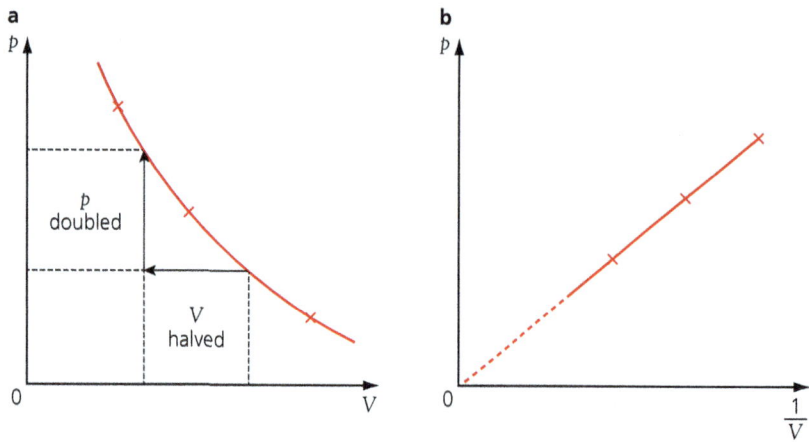

Figure 4.8 Graphical representations of Boyle's law

> **Examiner guidance**
>
> If you think pressure and volume are related by a power law (for example, $p \propto V$, $p \propto V^2$, $p \propto V^{-1}$, $p \propto V^{-2}$), write the relationship as $p \propto V^n$, or $p = kV^n$, where k is a constant. Taking logarithms (base 10 or natural logarithms, base e) gives $\log p = \log k + n \log V$. The gradient of the graph $\log p$ (y-axis) versus $\log V$ (x-axis) is n, the power required (in this case, –1).

■ ACTIVITY

3 Table 4.3 shows a set of pressure and volume readings for a fixed mass of ideal gas at constant temperature.

Pressure, p/kPa	500	245	170	125	100	50
Volume of gas, V/cm^3	1.00	2.00	3.00	4.00	5.00	10.00

Table 4.3 The results of a Boyle's law investigation

a Determine the value of $p \times V$ for each pair of readings.
b Plot a graph of pressure against the reciprocal of volume.
c Plot a graph of $\log p$ (y-axis) versus $\log V$ (x-axis) to verify that, in the relationship $p = kV^n$, n is approximately –1.

■ The pressure law

The apparatus in Figure 4.9 can be used to verify the pressure law: the relationship between pressure and temperature of a fixed mass of gas at constant volume.

> **Expert tip**
>
> The rubber tubing from the flask to the pressure gauge should be as short as possible. The flask must be in water almost to the top of its neck and be securely clamped to keep it off the bottom of the can.

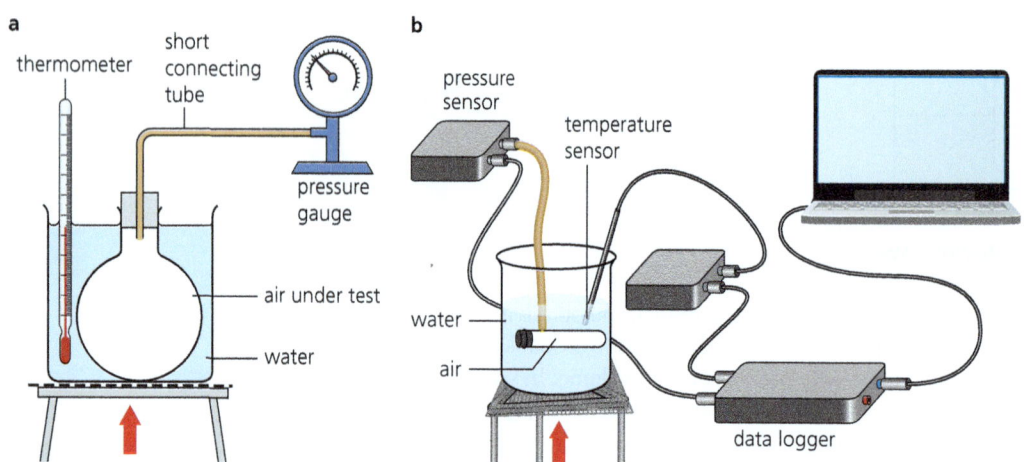

Figure 4.9 Pressure law apparatus

The procedure using the apparatus in Figure 4.9a is to record the pressure over a wide range of temperatures. Before recording a pressure reading, it is necessary to stop heating, stir the water and allow time for the pressure gauge reading to become steady; the air in the flask will then be at the temperature of the water (there will be thermal equilibrium).

A graph is plotted of pressure on the y-axis versus absolute temperature (in kelvin) on the x-axis (Figure 4.10). A linear relationship confirms the pressure law: the pressure is directly proportional to the absolute temperature, $p \propto T$, where T represents the absolute temperature in kelvin.

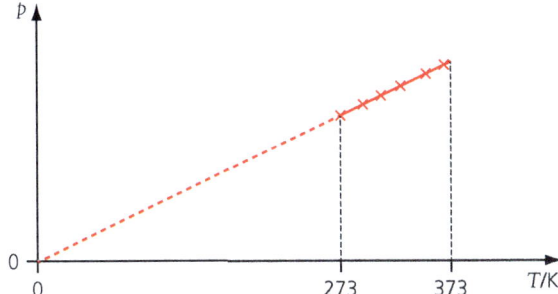

Figure 4.10 Graphical representation of the pressure law

> **Examiner guidance**
>
> If you assume that the gas behaves ideally at lower temperatures and the linear relationship continues to hold, then you can **extrapolate** the graph and determine an approximate value for absolute zero in degrees Celsius. If you have access to liquid nitrogen it is possible to get a reading at a much lower temperature to improve your line of best fit. This leads to a much lower degree of uncertainty.

> **Key definition**
>
> **Extrapolation** – Estimation of a value for a variable outside the range of the data, by assuming that the relationship between the variables continues to be valid. This is often done using a line graph by extending the line (or curve).

■ Charles' law

The apparatus to verify Charles' law, relating the volume and temperature of a fixed mass of gas at constant pressure, is shown in Figure 4.11.

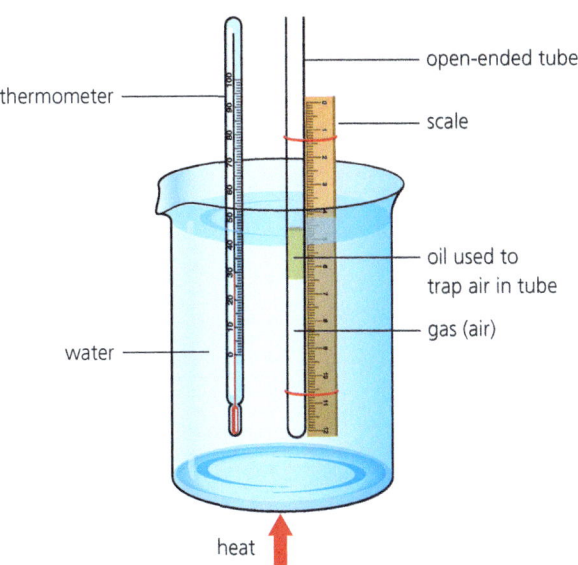

Figure 4.11 Charles' law apparatus

The relationship found is shown in Figure 4.12. This confirms Charles' law: volume $V \propto T$, where T represents the absolute temperature in kelvin.

An experimental value for absolute zero can be estimated by extrapolating the graph line and converting the absolute temperature in kelvin to degrees Celsius.

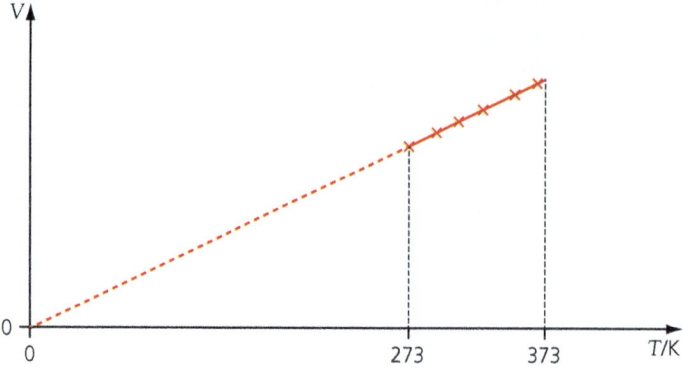

Figure 4.12 Graphical representation of Charles' law

ACTIVITY

4 Complete Table 4.4 and plot a graph of volume of gas (y-axis) versus temperature (x-axis). From your graph, estimate absolute zero.

Temperature, θ/°C	Absolute temperature, T/K	Volume of gas, V/dm^3	$\frac{V}{T}$/m^3 K^{-1}
273		0.1094	
100		0.0748	
10		0.0568	
1		0.0545	
0		0.0544	
−20		0.0503	

Table 4.4 Results of a Charles' law investigation

Determining the speed of sound using an electronic timer

■ Essential theory

When the surface of a solid vibrates, it will disturb the surrounding air and produce a series of compressions and rarefactions (variations in air pressure) that travel away from the surface as a longitudinal wave. If the waves have an audible frequency (in a range that can be detected by our ears), then they are described as sound. Since speed = $\frac{\text{distance}}{\text{time taken}}$, in order to determine the speed of sound it is necessary to measure the time it takes a sound wave to travel a known distance. Because the speed of sound is large (> 300 m s^{-1}), to obtain accurate results it is often necessary to use large distances and measure short time intervals accurately and precisely.

■ Suggested method

Two microphones are set about 1 metre apart; one is attached to the 'start' terminal and the other to the 'stop' terminal of an electronic timer, as shown in Figure 4.13. The timer must have millisecond accuracy. Measure and record the distance, d, between the centres of the microphones with a metre rule. With a small hammer and metal plate to one side of the 'start' microphone, produce a short, sharp sound (to ensure that the microphones send clear signals to the electronic timer).

When the sound reaches the 'start' microphone, the timer will start; when it reaches the 'stop' microphone, the timer will stop. The time displayed is then the time taken for the sound to travel the distance, d. Record the time and then re-set the timer; repeat the experiment several times and calculate a mean value for t for consistent results.

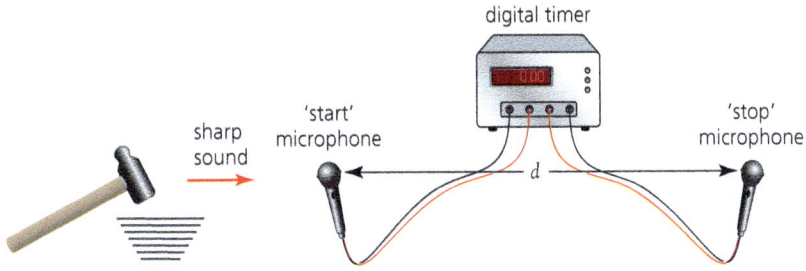

Figure 4.13 Measuring the speed of sound

RESOURCES

Experiments investigating the speed of sound:
- https://faraday.physics.utoronto.ca/IYearLab/solidsd.pdf
- http://www.pdst.ie/sites/default/files/ph_pr_soundexperiments.pdf
- https://www.sheffield.ac.uk/polopoly_fs/1.14268!/file/G5_speed-of-sound_V1.0.pdf

Resonance and speed of sound

Air columns in pipes or tubes of fixed lengths can vibrate with particular resonant frequencies. An example is an organ pipe of length, l, with one end closed. The air in the column, when driven at particular frequencies, vibrates in resonance.

The interference of the waves travelling down the pipe and the reflected waves travelling up the tube produces longitudinal standing waves, which have a node at the closed end of the pipe and an antinode at the open end.

The resonant frequencies of a pipe or tube depend on its length, l. Only a certain number of wavelengths can 'fit' into the pipe length with the node–antinode requirements needed at resonance.

Resonance occurs when the length of the pipe is equal to an odd number of quarter wavelengths, that is $l = \frac{\lambda}{4}$, $l = \frac{3\lambda}{4}$, $l = \frac{5\lambda}{4}$, or generally $l = \frac{n\lambda}{4}$ with $n = 1, 3, 5, ...$, and hence $\lambda = \frac{4l}{n}$. Incorporating the frequency, f, and the speed, c, through the wave equation $c = \lambda f$, or $f = \frac{c}{\lambda}$, we have $f = \frac{nc}{4l}$ with $n = 1, 3, 5, ...$. (Figure 4.14).

> **Expert tip**
>
> The accepted speed of sound at atmospheric pressure and 0 °C is 331.5 m s^{-1}. The speed of sound increases 0.607 m s^{-1} for every 1 °C rise in air temperature. The speed is directly proportional to the square root of absolute temperature and inversely proportional to the square root of the molar mass of the gas molecules.

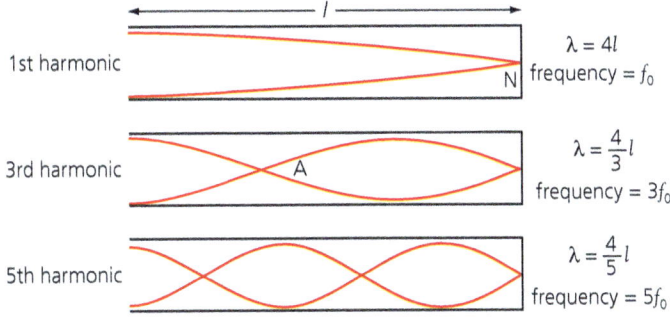

Figure 4.14 The first three possible harmonics in a pipe open at one end

Hence, an air column of length l has specific resonant frequencies and will be in resonance with the corresponding odd-harmonic driving frequencies. This can be made use of in an experiment to determine the speed of sound in air, as shown in Figure 4.15. As the length of the air column is increased, more wavelength segments will fit into the tube. When an antinode is at the open end of the tube, a loud resonance tone is heard. Lowering the water level in the tube and listening for successive resonances allows measurement of the tube lengths for antinodes to be at the open end of the tube. The difference in tube lengths when successive antinodes are at the open end of the tube is equal to a half wavelength.

■ **ACTIVITY**

5 Find out how Kundt's dust tube experiment can be used to find the speed of sound in air.

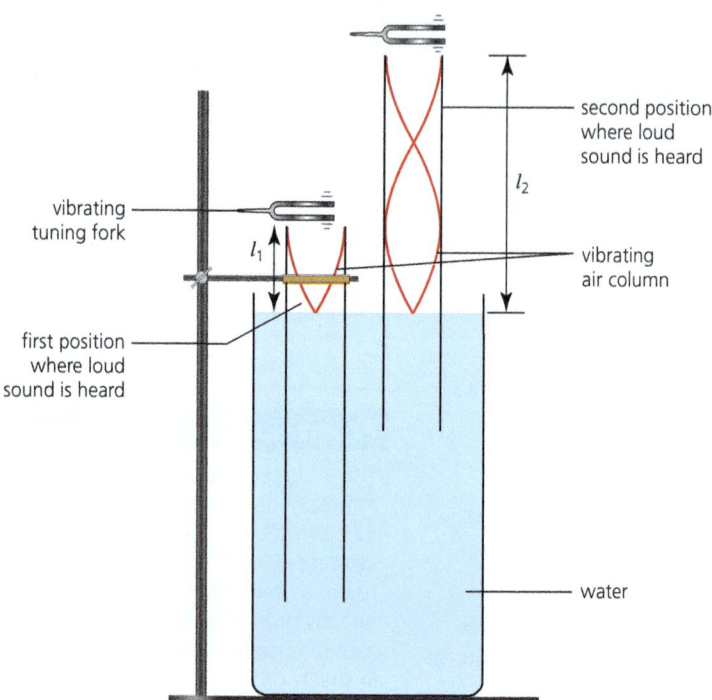

Figure 4.15 Using standing waves in an air column to determine the speed of sound in air

If the frequency, f, of the driving tuning fork is known and the wavelength is determined by measuring the difference in tube length between successive antinodes, $\Delta l = \frac{\lambda}{2}$ or $\lambda = 2\Delta l$. Then the speed of sound in air, c, can be calculated from the wave equation $c = \lambda f$.

> **Worked example**
>
> A student strikes one end of a 20.0 cm length of plastic pipe (a tube with one end closed) with a piece of metal and, using an oscilloscope, records the sound which it produces. It completes 10 cycles in 24 milliseconds.
>
> Determine the frequency f_0 of the sound.
>
> $f_0 = \dfrac{10}{0.024} = 417$ Hz
>
> Determine the wavelength, λ_0, assuming the sound produced is the fundamental resonance (first harmonic).
>
> $\lambda_0 = 4l = 0.80$ m
>
> Determine the frequencies of the next two harmonics.
>
> For the third harmonic:
>
> $f = 3f_0 = 1250$ Hz
>
> For the fifth harmonic:
>
> $f = 5f_0 = 2080$ Hz
>
> Deduce the speed of sound.
>
> $c = \lambda_0 \times f_0 = 333$ m s^{-1}

> **Expert tip**
>
> The antinode of a standing sound wave at the open end of a resonance tube is known to actually be located beyond the end of the tube, by a distance of about $0.61R$, where R is the radius of the tube. This 'end-correction' should be applied at each open end (but in the method shown in Figure 4.15 it is eliminated).

Determining refractive index

As a ray of light passes from a less optically dense to a more optically dense material, it bends (refracts) towards the perpendicular to the plane of the surface of the material (that is, the normal). Light entering a less optically dense material will refract (bend away) from the normal.

The refractive index, n, of a transparent solid, for example, glass, may be found by direct measurement of angles of incidence and refraction (Figure 4.16) using a ray box (or laser) in a darkened laboratory. Snell's law is used to determine a value for the refractive index; for light travelling from medium 1 to medium 2, the ratio of refractive indices is:

$$\frac{n_1}{n_2} = \frac{\sin\theta_2}{\sin\theta_1} = \frac{v_2}{v_1}$$

where θ denotes the angle with the normal and v denotes the speed of the light. When medium 1 is air, this results in:

$$n_2 = \frac{\sin\theta_1}{\sin\theta_2}$$

A graph of $\sin\theta_1$ versus $\sin\theta_2$ results in a straight line passing through the origin with the gradient equal to the refractive index of medium 2.

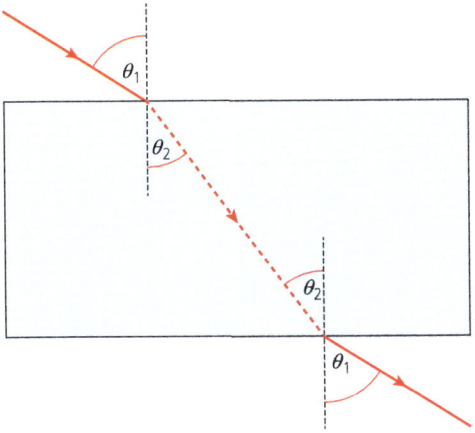

Figure 4.16 Investigating refraction in a glass block

Examiner guidance

The refractive index of air is almost that of a vacuum, which is equal to 1, so light travelling in air can be considered to be in a vacuum.

Expert tip

- Avoid using small angles of incidence as this will result in large percentage errors.
- Place two dots far apart on the incident and refracted light beams to accurately locate the beams.

■ ACTIVITY

6 Complete Table 4.5 which shows data from light incident on a glass block. Use the data to verify Snell's law.

Angle of incidence, θ_1/°	Angle of refraction, θ_2/°	$\sin\theta_1$	$\sin\theta_2$	$\frac{\sin\theta_1}{\sin\theta_2}$
30	19			
45	28			
65	37			

Table 4.5 Results from a refraction experiment

Worked example

Use the following data from a refraction experiment to determine:

a n_2

b the percentage errors in $\sin\theta_1$ and $\sin\theta_2$

c the uncertainty in n_2.

$n_1 = 1.000$

$\theta_1 = (22.03 \pm 0.80)°$

$\theta_2 = (10.45 \pm 0.80)°$

$n_2 = \frac{\sin\theta_1}{\sin\theta_2} = 2.07$

Percentage error in $\sin\theta_1 = \pm 3.6\%$

Percentage error in $\sin\theta_2 = \pm 7.7\%$

Uncertainty in $n_2 = \pm 11.3\% = 0.23$

RESOURCES

https://phet.colorado.edu/en/simulation/bending-light

Critical angle and total internal reflection

When a light ray in a glass block reaches the surface, it bends away from the normal as it refracts into the air. If the angle of incidence on the boundary is increased, when it reaches the critical angle the light no longer refracts outward, but instead travels along the surface (red ray in Figure 4.17). The angle of refraction is now 90°. The green ray shows total internal reflection.

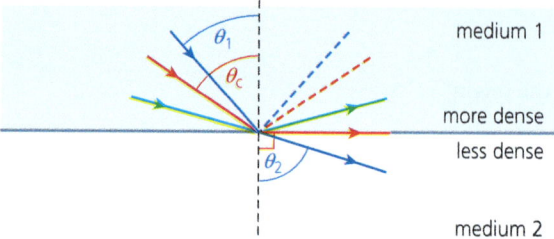

Figure 4.17 Total internal reflection occurs if the angle of incidence θ_1 is greater than the critical angle θ_c

The refractive index of a transparent medium, such as glass, can be determined experimentally using the relationship $n = \frac{1}{\sin \theta_c}$, where θ_c is the critical angle.

ACTIVITY

7   a   The critical angle for a certain type of glass is 42°. Calculate the refractive index of the glass.

   b   Determine the critical angle of a sample of glass whose refractive index is 1.46.

Real and apparent depth

Viewed from above, water looks shallower than it actually is due to the refraction of light at its surface. The ratio of the real depth to the apparent depth is the refractive index of the water (relative to air):

$$\text{refractive index} = \frac{\text{real depth}}{\text{apparent depth}}$$

The same relation holds for depths in other transparent media, when viewed in air. It is applied in an approach to measure the refractive index of glass (Figure 4.18). The travelling microscope is first focused on a cross drawn on the paper without the glass block. Record the reading on the micrometer scale, d_1. Place the glass block on the paper, refocus on the cross and record the reading on the micrometer scale, d_2. Mark a cross on top of the glass block and focus the travelling microscope on it; record the reading on the micrometer scale, d_3.

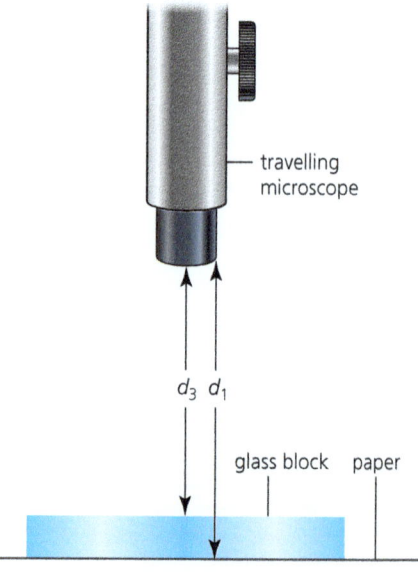

Figure 4.18 Determining the refractive index of glass (using a travelling microscope)

Expert tip

There will be a small percentage error in the micrometer readings, but a larger error may arise in judging the point at which the cross is in focus. Establish a range over which it can be considered to be in focus and estimate the error in the reading from this range.

RESOURCES

Investigating Newton's rings:
www.schoolphysics.co.uk/age16–19/Wave%20properties/Interference/text/Newton%27s_rings/index.html

The refractive index of the glass is calculated from the expression:

$$\text{refractive index} = \frac{d_3 - d_1}{d_3 - d_2}$$

Investigating factors that determine resistance

■ Essential theory

Ohm's law states that for a conductor the current, I, is directly proportional to the potential difference, V, provided physical factors such as temperature and pressure remain constant.

At a constant temperature the resistance of a wire of a particular metal depends on the length of the wire and its cross-sectional area (determined by its thickness).

■ Suggested method

One method of measuring electrical resistance is to measure the voltage drop, V, across a resistance in a circuit with a voltmeter and the current, I, through the resistance with an ammeter. Then, providing Ohm's law holds, the resistance is $R = \frac{V}{I}$.

This is known as the ammeter–voltmeter method and is shown in Figure 4.19 for a wire. This simple apparatus can be used to investigate the effect of length and cross-sectional area for a range of different metals. The movable crocodile clip should be pressed firmly on to the cleaned wire to minimize contact resistance. Readings from the ammeter and voltmeter are recorded with the crocodile clip in different positions to establish the relationship between resistance and length. The procedure can be repeated with different wires but a fixed distance between the clips.

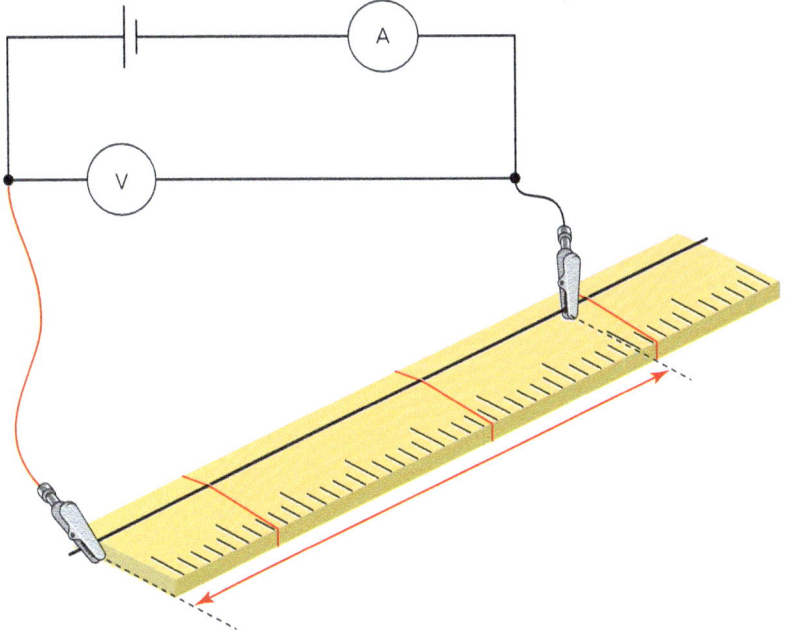

Figure 4.19 Measuring the resistance of different lengths of wire

> **Examiner guidance**
>
> The wire normally gets hot during this experiment which raises the temperature and resistance of the wire. It is best to turn off the power supply between readings so as to avoid the wire heating up.

■ ACTIVITY

8 Table 4.6 shows the results of an experiment into the effect of length on the resistance of nichrome wire of diameter 0.45 mm. Plot a graph of resistance, R, on the y-axis versus length, l, on the x-axis. State and explain the observed relationship.

Length of wire, l/m	0.00	0.25	0.50	0.75	1.00
Resistance of wire, R/Ω	0.00	0.60	1.20	1.90	2.50

Table 4.6 Results from an investigation into the relationship between length of wire and its electrical resistance

9 Table 4.7 shows the results of an experiment into the effect of cross-sectional area on the resistance of nichrome wire, using wires of different diameters but fixed length of 0.50 m. Plot a graph of resistance, R, on the y-axis versus cross-sectional area, A, on the x-axis. State and explain the observed relationship.

Diameter, d/mm	0.45	0.57	0.71	0.91
Cross-sectional area, A/m^2	1.6×10^{-7}	2.6×10^{-7}	4.0×10^{-7}	6.5×10^{-7}
Resistance, R/Ω	3.1	2.0	1.3	0.8

Table 4.7 Results from an investigation into the relationship between cross-sectional area of wire and its electrical resistance

RESOURCES

- https://phet.colorado.edu/en/simulation/resistance-in-a-wire
- https://phet.colorado.edu/en/simulation/ohms-law

Expert tip

The ratio of the measured voltage and current does not give an exact value of the resistance because of the resistance of the meters. Ideally, the voltmeter should have infinite resistance while the ammeter should have zero resistance. It is wise to measure the voltage and current separately to avoid possible errors recording the two measurements together.

Examiner guidance

Resistance is proportional to length l and inversely proportional to cross-sectional area A.

$$R = \text{constant} \times \frac{\text{length}}{\text{cross-sectional area}}$$

The constant is a property of the material used, known as its resistivity, ρ (unit Ω m).

$$R = \frac{\rho l}{A}$$

Determining internal resistance

■ Essential theory

A cell or battery has an internal resistance. Electrons leaving the negative terminal will be involved in collisions within the material of the cell itself and will experience electrical resistance.

Some of the electrical energy that the emf supplies to the moving charge is used in overcoming the cell's internal resistance and so the circuit components receive a slightly lower voltage than if there was no internal resistance. The 'terminal voltage' (pd across the cell terminals) is less than the emf of the battery, and as the current increases the difference becomes greater.

Thermal energy is dissipated in the cell due to the internal resistance which may make it hot when in use.

Figure 4.20 shows the emf, ε, internal resistance, r, current, I and a resistor, R, in series with the cell.

Note that the current, I, will be the same in both resistances because they are in series. According to Kirchhoff's second law, the emf will be shared across r and R.

So, as $V = IR$,

$\varepsilon = IR + Ir$ or $\varepsilon = I(R + r)$

The term Ir is referred to as the 'lost volts' because it is not available to the rest of the circuit. IR is the terminal pd, the voltage across the terminals of the cell or battery when the current, I, is being drawn. This is also the pd, V, measured across R, so $V = \varepsilon - Ir$.

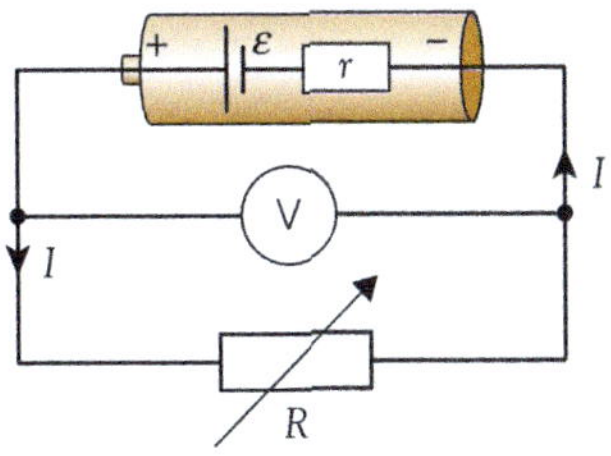

Figure 4.20 The concept and measurement of internal resistance

■ Suggested method

The emf, ε, could be measured directly with a multimeter or ideally with an oscilloscope, which has a high resistance, assumed to be infinite. This means there is very little or zero current in the circuit and hence the voltage measured on the oscilloscope is equal to ε.

Alternatively, a graphical approach can be used that allows you to work out the internal resistance, r, as well as the supply emf, ε. If, in the circuit in Figure 4.20, R is replaced with a variable resistor, a set of values can be obtained for V and I using the voltmeter and an ammeter. A straight line ($y = mx + c$) is obtained for a plot of voltage, V, versus current, I, with a gradient of $-r$ and a y-intercept of ε (Figure 4.21).

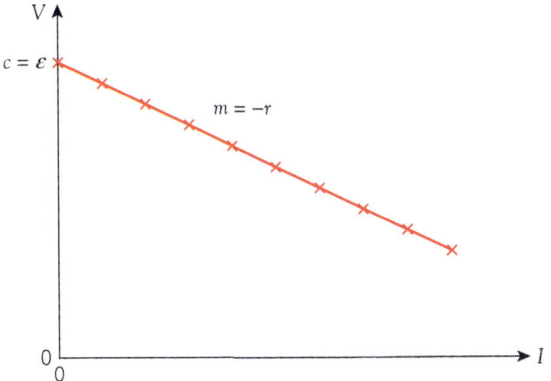

Figure 4.21 A graph of voltage, V, against current, I, showing the determination of the internal resistance, r

Worked example
A high-resistance voltmeter is connected across an AA cell and reads 1.56 V. A 4.70 Ω resistor is then connected across the cell and the voltmeter reading drops to 1.48 V. Calculate the internal resistance of the cell. $\varepsilon = IR + Ir$ Putting $I = \dfrac{V}{R}$ gives $\varepsilon = V + \dfrac{Vr}{R}$ which leads to: $r = R \times \left(\dfrac{\varepsilon}{V} - 1\right)$ $r = 4.7\,\Omega \times \left(\dfrac{1.56V}{1.48V} - 1\right) = 0.25\,\Omega$

RESOURCES

Investigating internal resistance:
- https://www.youtube.com/watch?v=aWIB6ZMzwHw
- http://www.nuffieldfoundation.org/practical-physics/internal-resistance-shoe-box-cell
- http://practicalphysics.org/internal-resistance-potato-cell.html

Investigating half-life (experimentally or via simulation)

■ Essential theory

Radioactive decay (or disintegration) is due to unstable nuclei, which emit α-particles, β-particles or γ-rays and become more stable (Figure 4.22). It is a spontaneous and random process. When a nucleus decays, it changes into a nucleus of another element.

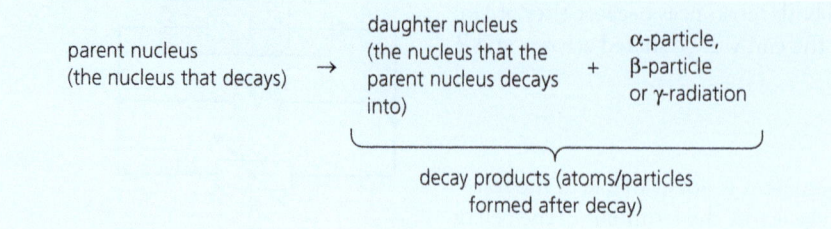

Figure 4.22 Radioactive decay processes

Table 4.8 summarizes the three types of radioactive decay.

Type	Nuclear equation	Nuclear process
α decay	$^{A}_{Z}X \rightarrow\ ^{A-4}_{Z-2}Y +\ ^{4}_{2}He$	A helium-4 nucleus (α-particle) is released from the parent nucleus.
β decay	$^{A}_{Z}X \rightarrow\ ^{A}_{Z+1}Y +\ ^{0}_{1}e$	One of the neutrons in the parent nucleus changes into a proton and an electron (β-particle) (and an anti-neutrino).
γ emission	$^{A}_{Z}Y \rightarrow\ ^{A}_{Z}Y + \gamma$	After emitting an α- or a β-particle, some nuclei are left with more energy than normal. This extra energy is emitted as γ-rays to make the nuclei more stable.

Table 4.8 Alpha, beta and gamma decay

The activity of the radioactive material is the number of disintegrations per second, measured in becquerel (Bq). The half-life of a radioactive nuclide is the time taken for half of the nuclei present in any given sample to decay. It is also the time taken for the activity of a given sample to fall to half of its original value. The number of undecayed nuclei and hence the activity halves at regular intervals; in other words, the half-life of a radioactive source is a constant.

■ Suggested method

The isotope protactinium-234, ^{234}Pa, has a half-life of several tens of seconds. With suitable safety precautions, you can monitor its decay every 10 seconds using a GM tube (see Chapter 3) connected to a ratemeter/scaler or to a data-logging interface and computer, and hence determine its half-life experimentally (Figure 4.23).

You must consult your physics teacher before working with radioactive isotopes.

The sample of ^{234}Pa is produced by the decay of ^{238}U. (^{234}Pa is the granddaughter of ^{238}U, which decays via alpha decay to ^{234}Th then via beta decay to ^{234}Pa). The ^{238}U is in the form of yellow uranyl nitrate $(UO_2(NO_3)_2)$ solution in water and is contained in a sealed plastic bottle. The bottle also contains an oily solvent that floats above the water.

When the bottle is shaken, some of the ^{234}Pa in the aqueous layer dissolves into the oily layer. Once the two layers have separated out, ^{234}Pa stops moving into the oily layer, so there is a fixed amount of ^{234}Pa in the oily layer. ^{234}Pa emits high-energy beta radiation, which can penetrate the plastic bottle and travel some distance in air.

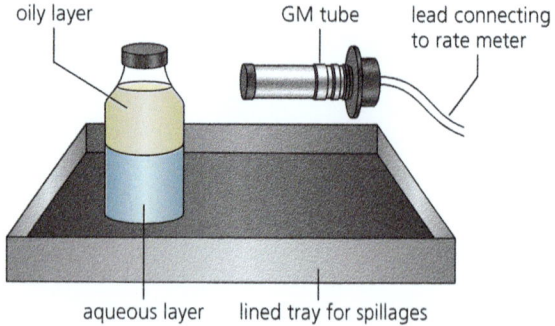

Figure 4.23 Measuring the half-life of protactinium-234

A graph of count rate (taking into account the background count rate, see Chapter 3) versus time is plotted to establish the half-life of protactinium-234.

> **Examiner guidance**
>
> When discussing the activity of a source you need to include an efficiency factor, which could be small (for example, 1%). This is because the GM tube detects only a fraction of the decays. The detected count rate is not in Bq, unless it is corrected for efficiency.

■ ACTIVITY

10 The activity A of a radioactive source is given by $A = A_0 e^{\lambda t}$, where A_0 is the activity at time $t = 0$ and λ is the decay constant. Table 4.9 shows the results of detecting the count rate from a radioactive source. The count rates tabulated have been corrected for background radiation.

Time, t/s	Count rate, A
30	5768
150	3391
270	1963
390	1231
510	718
630	415

Table 4.9 Detected count rate from a radioactive source

a Plot a linear graph showing the variation of the count rate against time.
b Plot a suitable semilog graph to determine λ and A_0.

> **RESOURCES**
>
> Investigating radioactivity:
> - http://zenit.winbasonline.se/productfiles/871_19-093483.pdf
> - https://www.youtube.com/watch?v=Ir5R_0KQBsk
> - https://phet.colorado.edu/en/simulation/legacy/beta-decay

Investigating Young's double-slit experiment (HL only)

■ Essential theory

The wavelength of monochromatic light can be determined by using a double slit to produce an interference pattern, known as Young's fringes (Figure 4.24).

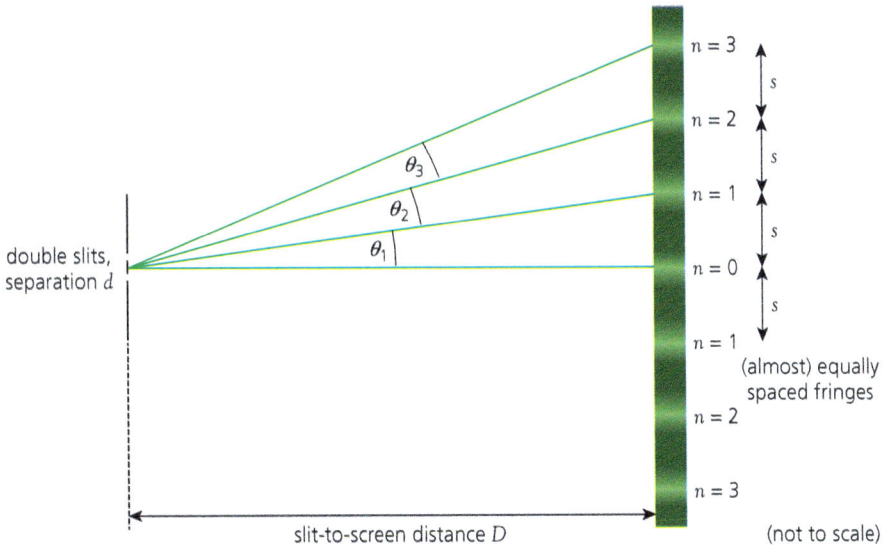

Figure 4.24 Separation and numbering of fringes seen on a screen in a Young's double slit experiment

Figure 4.25 shows the relationship between the various quantities in this arrangement.

The light waves leaving each slit act as two separate beams of light with the same wavelength and (provided they originate from the same suitable source) leave the slits in phase with one another – they are *coherent*.

Where the two light waves have travelled the same distance from both slits to the screen (that is, in the middle of the interference pattern), they will still be in phase so will interfere constructively and there will be a bright fringe. This central maximum is the brightest of all the fringes.

Moving to either side of the central fringe, there will be another bright fringe at the point where the wave from one of the slits is once again in phase with the wave from the other slit. This occurs when the wave from the 'further' of the two slits has travelled exactly one wavelength more than the other: the path difference is equal to one wavelength.

> **Expert tip**
>
> Coherence, that is having the same frequency and a constant phase relationship, is necessary to produce a clear interference pattern.

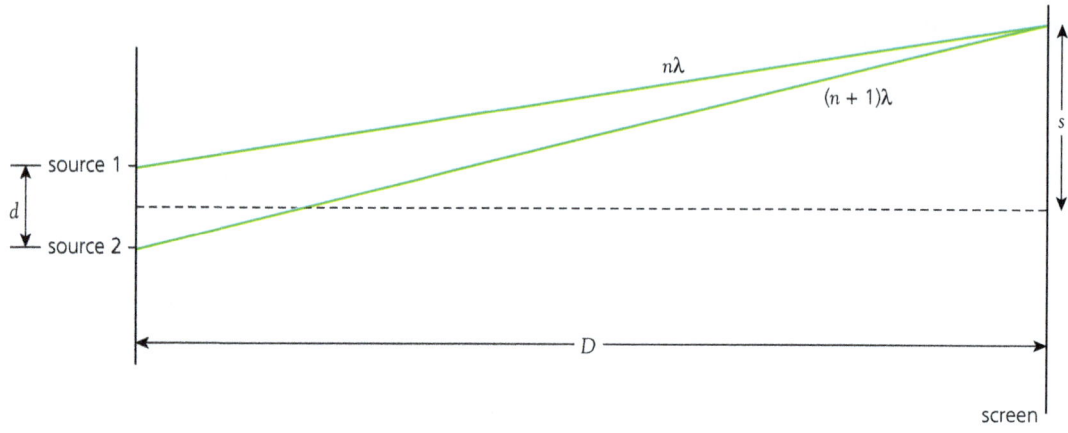

Figure 4.25 Path difference for a bright fringe in double-slit interference

There will be further bright fringes at each position on the screen where the path difference is exactly equal to two wavelengths, three wavelengths, and so on. The bright fringes will be almost equally spaced on the screen.

In between these bright fringes will be 'dark fringes' formed by destructive interference. At each of these positions, the path difference will be equal to a $\frac{1}{2}$ wavelength, $1\frac{1}{2}$ wavelengths, $2\frac{1}{2}$ wavelengths, and so on, resulting in the light waves being 180° out of phase.

The equation
$$s = \frac{\lambda D}{d}$$
gives the separation of the fringes for slit separation, d, and slit-to-screen distance, D (see Figure 4.25).

Examiner guidance

Remember to convert all distances into metres before substituting into this equation. Visible light wavelengths are usually quoted in nanometres (nm).

Worked example

Monochromatic light falling on two narrow slits 0.20 mm apart produces an interference pattern on a screen 2.00 m away with bright bands 5.00 mm apart. Calculate the wavelength of the light and deduce in which part of the visible spectrum it lies.

$$\lambda = \frac{5 \times 10^{-3}\,\text{m} \times 2 \times 10^{-4}\,\text{m}}{2.00\,\text{m}} = 5 \times 10^{-7}\,\text{m}$$

This is in the blue–green region of the visible spectrum.

Suggested method

Young's double-slit experiment involves a slide with two narrow, parallel slits very close together. A laser beam is incident on both slits, passes through each and spreads out slightly by diffraction (Figure 4.26).

> **Expert tip**
>
> The light from a laser is coherent: all of the light waves emitted have the same frequency and a constant phase relationship.

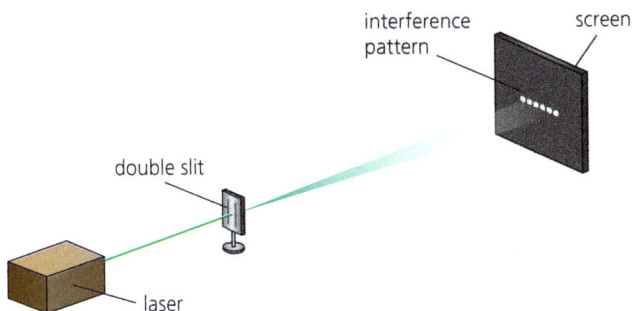

Figure 4.26 Apparatus for demonstrating interference of light waves

A screen is placed several metres distance from the laser (in a darkened lab) and interference fringes can be seen.

By measuring the split spacing, d, the fringe spacing, s, and the slit-to-screen distance, D, the wavelength, λ, of the light can be determined from the expression $s = \frac{\lambda d}{D}$. A graph of d on the y-axis against D on the x-axis should be a straight line through the origin, with gradient $\frac{\lambda}{s}$.

> **Examiner guidance**
>
> You should measure across several fringes and obtain the fringe spacing by dividing this measurement by the number of fringes. It may be easier to measure from the centres of the dark fringes, as these may be easier to locate. The uncertainty of a ruler (±1 mm) applies to the end points of the measurement – so if you measure 10 fringe spacings as 120 mm ± 1 mm, a single fringe spacing is 12.0 mm ± 0.1 mm, since it has the same percentage uncertainty.

■ ACTIVITY

11 A Young's double-slit experiment was set up to produce interference fringes on a screen using a monochromatic source of green light. The fringes were too close together for measurement and observation. State the effects of the following changes on the separation of the fringes:

   a decreasing the distance between the screen and slits
   b increasing the distance between the source and the slits
   c increasing the distance between the two slits
   d increasing the width of each slit (limited to interference effects only)
   e replacing the light source with a monochromatic source of red light.

Investigating a diode bridge rectification circuit (HL only)

■ Essential theory

Rectification is the process in which an alternating current (ac) is forced to flow only in one direction (that is, it is converted to dc). It is achieved using diodes (semiconducting components) because these allow current to flow in one direction only. They have low resistance in one direction (*forward bias*), but very high resistance in the opposite direction (*reverse bias*).

Suggested method

Full-wave rectification is achieved by the setup in Figure 4.27 which uses four diodes arranged into a bridge rectifier circuit. The four diodes are connected in a diamond-shaped pattern and the input terminals are P and Q. If P is positive during the first half-cycle of the ac input, rectifiers 1 and 2 will conduct and current will flow through the local resistor from + to −. In the next half-cycle, Q is positive and rectifiers 3 and 4 will conduct; current will still flow through the local resistor from + to −. Therefore the resistor will always have its upper terminal positive and its lower terminal negative: dc current will always flow through it in the same direction.

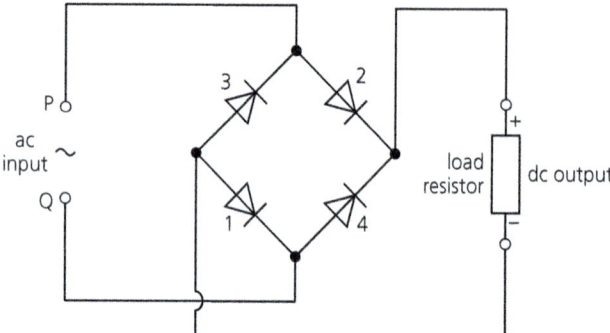

Figure 4.27 Four-diode (bridge) circuit for full-wave rectification

To smooth the varying dc output (the broken curve in Figure 4.28), a capacitor is connected across the output so that it charges up from the rectified pd and then slowly (depending on the 'time constant', RC, where R denotes the load resistance in the smoothing circuit) releases electrons to a load. The time constant delays the change in the output. If the time constant is large the voltage across the load will not have fallen significantly before the next half-cycle of the rectified pd arrives to rapidly restore the charge on the capacitor back to its peak value.

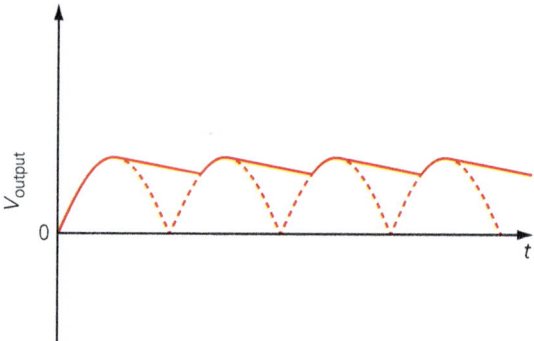

Figure 4.28 Smoothing of output voltage by a capacitor

The variation in voltage in Figure 4.28 can be observed by connecting an oscilloscope across the load resistor. You should draw waveforms and record the values of R (R denoting the load resistance in the smoothing circuit) and C and the oscilloscope settings. You could investigate the effect of various positions for the capacitor and introducing an inductor.

You could also investigate the major differences between a full-wave rectification circuit with four-diode bridge connection and another circuit in which two of the diodes are replaced with resistors.

> **RESOURCES**
>
> Simulations of electrical circuits including rectification:
> - http://www.falstad.com/circuit/
> - http://www.falstad.com/circuit/e-fullrect.html

Investigating a compound microscope (Option C only)

Essential theory

The compound microscope is a system of two converging lenses, used to look at very small objects at short distances. The lens closest to the object, the objective, is used to enlarge and invert the object into a 'real' image. The lens closest to the eye, the eyepiece, acts as a simple magnifier, used to view the image formed by the objective. The simple magnifier is a converging lens placed close to the eye that increases the size of the image formed on the retina. Figure 4.29 shows a ray diagram for the compound microscope.

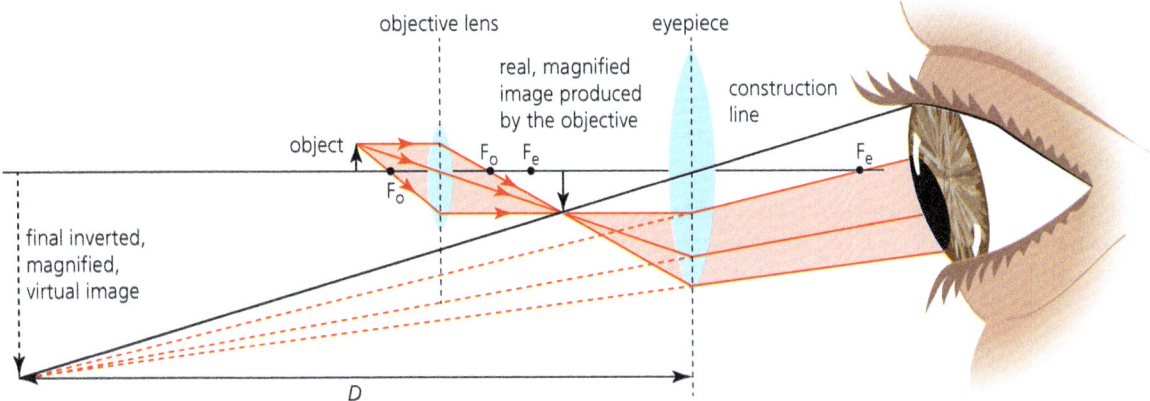

Figure 4.29 Compound microscope with the final image at the near point (normal adjustment); the size of the lenses and their separation are *not* to scale

To locate the image by drawing you need to find the point where the construction line through the centre of the eyepiece from the top of the first image meets the extension of the ray from the first image that passes through the focal point of the eyepiece.

■ Suggested method

The object to be viewed under the microscope is placed just beyond the focal point of the objective lens, so that a real image is formed between the two lenses with a high magnification. The position of the eyepiece is adjusted to give as large an image as possible, with the final virtual image usually at, or very close to, the near point of the eye.

A model of a simple compound microscope can be investigated in a darkened lab, as shown in Figure 4.30. A converging lens with a focal length of about 5 cm is used to form an inverted image of a brightly illuminated object (for example, graph paper) on a translucent screen. Then the position of a second, less powerful lens is adjusted until a second (virtual) image of the first image is seen when looking through this eyepiece. The screen can then be removed and the two lenses used together to observe the scale, so that the magnification of the image can be estimated.

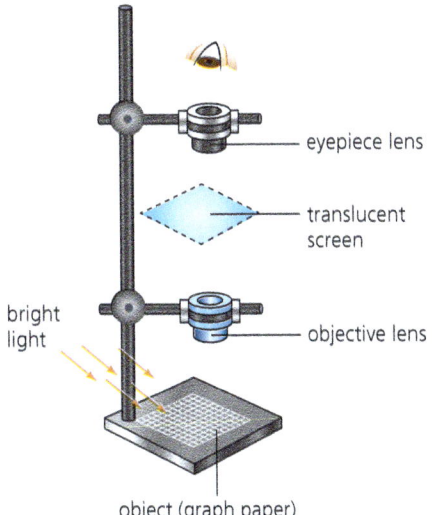

Figure 4.30 Investigating a model compound microscope

> **Examiner guidance**
>
> The angular magnification produced by a compound microscope is equal to the product of the linear magnification of the objective lens multiplied by the angular magnification of the eyepiece lens. For an image at the near point:
>
> $$M_{overall} = m_{objective} \times M_{eyepiece} = \left(\frac{-v}{u}\right) \text{objective} \times \left(\frac{D}{f} + 1\right) \text{eyepiece}$$
>
> D is the distance between the centre of the eyepiece lens and the final virtual image. If the final image is at infinity (for less eye strain), the +1 term can be omitted.
>
> The exact angular magnification of a microscope clearly depends on where the object and final image are located, but an approximation can be obtained from the focal lengths and the distance between the lenses, L:
>
> $$M \approx \frac{DL}{f_o f_e}$$

RESOURCES

Investigating a compound microscope:
- https://www.physast.uga.edu/~zhaoy/MillerTurnage.pdf
- http://hyperphysics.phy-astr.gsu.edu/hbase/geoopt/micros.html

Investigating a refracting telescope (Option C only)

■ Essential theory

An astronomical telescope is used to obtain magnified images of distant objects. Like the compound microscope, it is composed of two lenses, an objective and an eyepiece. Rays from a distant object will be focused by the objective to a point at or very near to its focal point. A real, inverted image is formed between the lenses (Figure 4.31).

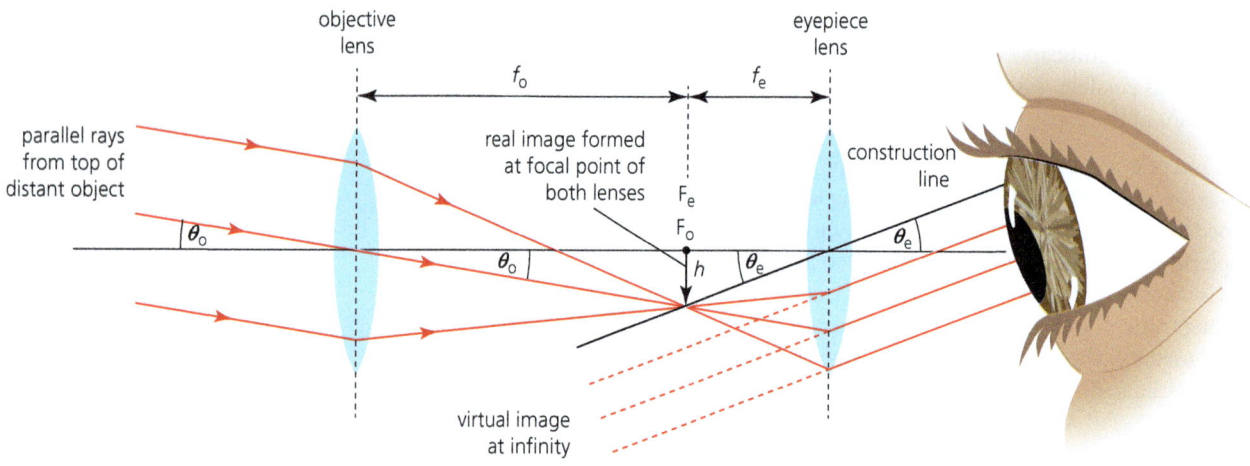

Figure 4.31 Simple refracting telescope with the final image at infinity (normal adjustment)

The telescope is designed so that the image from the objective lens forms at the focal point of the eyepiece, which acts as a simple magnifier. The final image (the image from the eyepiece) is enlarged, inverted and located at infinity. The distance between the two lenses is therefore equal to the sum of their focal lengths.

The direction to the top of the final image is located by drawing a construction line through the centre of the eyepiece from the top of the first image.

The total magnification of the reflecting telescope is derived from the angular magnification:

$$M = \frac{\theta_e}{\sigma_o} = \frac{\frac{h}{f_e}}{\frac{h}{f_o}}$$

$$M = \frac{f_o}{f_e}$$

A higher magnification is therefore obtained using an objective lens with a longer focal length (lower power) and an eyepiece lens with a smaller focal length (higher power). However, lens aberrations of high-power eyepieces limit the angular magnification.

■ Suggested method

A model of an astronomical telescope can be investigated in a darkened room as shown in Figure 4.32. A converging lens (focal length ≈ 50 cm) forms an inverted image of a brightly illuminated object (for example, a scale) on a translucent screen. The position of a second, more powerful lens (the eyepiece) is adjusted until a second virtual image of the first image is seen. The screen can then be removed and the two lenses used together to observe the scale, so that the magnification of the image can be estimated.

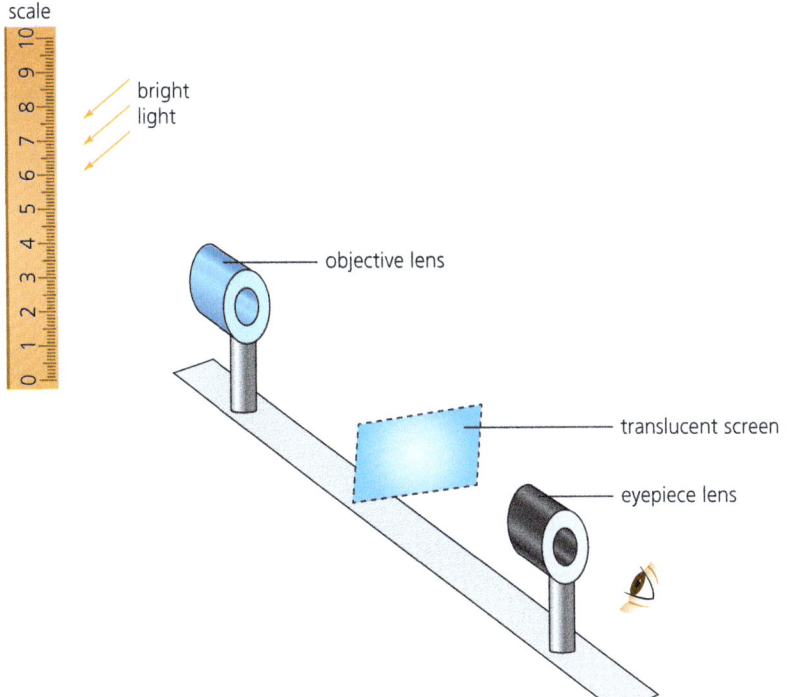

Figure 4.32 Investigating a model telescope

> **RESOURCES**
>
> Investigating a refracting telescope:
> https://wp.optics.arizona.edu/mnofziger/wp-content/uploads/sites/31/2016/05/OPTI202L-Lab-4-Telescopes-SP11.pdf

5 Mathematical and measurement skills

Scientific notation and significant figures

Very large and very small physical quantities are often written in **scientific notation**: $N \times 10^n$, where N is a number between 1 and 10, and n is the exponent (power) to which 10 is raised. For example:

6400 m = 6.4×10^3 m

0.000 0061 s = 6.1×10^{-6} s

■ ACTIVITY

1 a Write the following numbers in scientific notation:

    1002   54   6 926 300 000   −393   0.003 61   −0.0038

  b Write the following numbers in ordinary notation:

    1.93×10^3 3.052×10^1 -4.29×10^{-2}

    6.261×10^6 9.513×10^{-8} 9.512×10^{-11}

Scientific notation allows numbers to be expressed in a form that shows the number of **significant figures** (Table 5.1).

Value	Number of significant figures
1.3×10^4	2
4.0×10^3	2
5.70×10^3	3
3.600×10^6	4

Table 5.1 Selected numbers and the number of their significant figures

■ Relevance of significant figures in practical work

The word 'significant' in the mathematical sense indicates that the digit is 'meaningful'. Scientists report all the digits in a number that they know for certain plus one extra digit which is an estimate. All of these numbers, including the estimated number, are regarded as significant.

The number of significant figures (sf) in a data value therefore implicitly indicates the uncertainty in its measurement. The final figure in the number is unreliable. For example, 1.25 cm implies a minimum possible uncertainty of ±0.01 cm and a **range** of 1.24 cm to 1.26 cm.

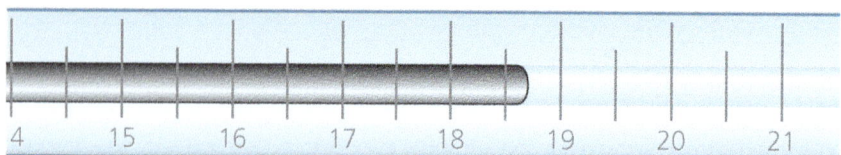

Figure 5.1 A magnified thermometer scale showing a temperature of 18.7 °C: the last digit is uncertain

Examiner guidance

There is therefore no need for you to describe a measurement as 'about…' or 'approximately…', or 'around…', because the precision of any measurement is incorporated into the number of digits you use to report that measurement. The magnitude of a reported uncertainty will provide an additional prompt regarding the precision of the value.

Expert tip

A shortcut for converting a number from ordinary decimal form to scientific notation is to move the decimal place a number of places that corresponds to the appropriate power of 10.

Key definitions

Scientific notation – A method for expressing a given quantity as a number between 1 and 10, having significant digits necessary for a specified degree of accuracy, multiplied by 10 to the appropriate power.

Significant figures – The digits of a number that are used to express it to the required degree of accuracy, starting from the first non-zero digit.

Range – The difference between the smallest and largest values.

RESOURCES

Scientific notation:
www.georgebrown.ca/uploadedFiles/TLC/_documents/Scientific%20Notation.pdf

You must report a measurement with the correct number of significant figures. For example, a mass reading for a metal block on a digital balance is 18.2 g ± 0.1 g (3 sf). A record of this mass as 18 g (2 sf) would have too few significant figures; the value does not reflect the sensitivity of the balance. A record of the mass as 18.20 g (4 sf) would have too many significant figures; the balance is not capable of reading to ±0.01 g, as the reading 18.20 g implies.

> **Expert tip**
>
> The precision of a measurement only equals the resolution of the measuring device if all repeated measurements are identical.

Establishing the number of significant figures

The following rules should be applied to establish the number of significant figures:

- Digits that are not zero are always significant. So, 343 m has 3 sf and 3.438 kg has 4 sf.
- Zeros that lie between non-zero digits are always significant. So, 6008 kg has 4 sf and 5.01 cm^3 has 3 sf.
- Zeros at the beginning of a number are never significant; they set the position of the decimal point. Hence, 0.334 g has 3 sf and 0.005 dm^3 has 1 sf.
- Zeros at the end of a number are always significant if the number contains a decimal point. Hence, 210.0 nm has 4 sf and 0.0700 kg m^{-3} has 3 sf.
- Zeros at the end of a number may or may not be significant if the number contains no decimal point.

The last point above means that 500 mm may have 1, 2 or 3 sf. It is impossible to decide which is correct without more information regarding how the measurement was recorded. This can be resolved by reporting the value in standard form or by stating the uncertainty in the value.

> **Examiner guidance**
>
> Placeholder zeros can be removed by converting numbers to scientific notation. For example, 2000 may have 1, 2, 3 or 4 sf, but 2.00×10^3 clearly has 3 sf.

ACTIVITY

2 State the number of significant figures in the following measurements:

   41 cm   14.00 g   0.0066 m   20.20 mol   −0.0038 °C   105.50 kg 5×10^3 J

Rounding off significant figures

A digit of 5 or larger rounds up; a digit smaller than 5 rounds down. When rounding look only at the one figure beyond the number of figures to which you are rounding. So, to round to 3 sf only look at the fourth figure.

> **Key definition**
>
> **Decimal places** – The number of digits, including zeros, to the right of the decimal point.

ACTIVITY

3 Round off each number to the required number of significant figures:

   a 1.2367 to 4 sf
   b 1.2384 to 4 sf
   c 0.01352 to 3 sf
   d 2.051 to 2 sf.

> **RESOURCES**
>
> An online significant figure calculator:
> www.calculatorsoup.com/calculators/math/significant-figures-rounding.php

Significant figures in simple calculations

When two or more measured values are added and/or subtracted, the final calculated value must have the same number of **decimal places** (dp) as the measured value that has the least number of dp.

For example:

$1.462\,36 \times 10^8 + 4.293 \times 10^7 = 1.462\,36 \times 10^8$ (5 dp) $+ 0.4293 \times 10^8$ (4 dp)
$= 1.891\,66 \times 10^8$ (uncorrected) $= 1.8917 \times 10^8$ (corrected to 4 dp)

> **Expert tip**
>
> Note that if you are adding numbers expressed in scientific notation then you need to adjust the numbers so they have the same exponent.

When two or more measured values are multiplied and/or divided, the final calculated value must have as many sf as the measured value that has the least number of sf. Table 5.2 gives worked examples of such calculations, showing incorrect and correct answers.

> **Examiner guidance**
>
> Some numbers are exact because they are known or defined with complete certainty. For example, there are exactly 60 seconds in 1 minute. You might need to do a calculation involving an exact number. For example, a simple pendulum makes exactly 10 complete oscillations in a time of 15.6 s. The period = 15.6 s ÷ 10 = 1.56 s. The time 15.6 s has 3 sf and 10 is an exact number, so the period calculated should have 3 sf.

Examples of multiplication	Explanations
1.2 × 3.45 = 4.14 (incorrect)	The number 1.2 has 2 sf while 3.45 has 3 sf.
1.2 × 3.45 = 4.1 (correct)	The answer will follow the number with the least sf. In this case, it is 2 sf.
	The answer should be rounded to 2 sf: 4.14 becomes 4.1.
0.0123 × 81.65 = 1.004 295 (incorrect)	The number 0.0123 has 3 sf, while 81.65 has 4 sf.
0.0123 × 81.65 = 1.00 (correct)	The answer will follow the least sf. In this case, it is 3 sf.
	The answer should be rounded to 3 sf: 1.004 295 becomes 1.00.
Examples of division	**Explanations**
0.87 ÷ 0.0344 = 25.290 70 (incorrect)	The number 0.87 has 2 sf while 0.0344 has 3 sf.
0.87 ÷ 0.0344 = 25 (correct)	The answer will follow the least sf. In this case, it is 2 sf.
	The answer should be rounded to 2 sf: 25.290 70 becomes 25.
0.0841 ÷ 1.2 = 0.070 0833 (incorrect)	The number 0.0841 has 3 sf while 1.2 has 2 sf.
0.0841 ÷ 1.2 = 0.070 (correct)	The answer will follow the least sf. In this case, it is 2 sf.
	The answer should be rounded to 2 sf: 0.070 0833 becomes 0.070.
1.21 ÷ 0.372 = 3.252 6882 (incorrect)	The number 1.21 has 3 sf and 0.372 also has 3 sf.
1.21 ÷ 0.372 = 3.25 (correct)	The answer should be rounded to 3 sf: 3.252 6882 becomes 3.25.

Table 5.2 Worked examples of calculations with measured values, showing incorrect and correct answers

For multiple calculations, first do logarithms, then exponents, then multiplication and division and finally addition and subtraction. When parentheses are used, do the operations inside the parentheses first. To avoid rounding errors, keep extra digits until the final step.

■ Logarithms, exponents and significant figures

In values of logarithms, retain in the mantissa (the number to the right of the decimal point) the same number of sf as there are in the number whose logarithm you are taking. For example, $\log_{10}(12.8) = 1.107$. The mantissa is .107 and has 3 sf because 12.8 has 3 sf.

In values of exponents, the number of sf is the same as in the mantissa of the power. For example, $10^{1.23} = 17$ or 1.7×10^1, which has 2 sf since there are 2 sf in the mantissa of the power (.23).

> ■ **ACTIVITY**
> 4 Calculate $\log_{10}(4.500 \times 10^3)$, $\log_{10}(4.50 \times 10^3)$ and $\log_{10}(4.5 \times 10^3)$ to the correct number of sf.

Processing raw data

It is usual for processed data to be given the same number of sf as raw data. Consider determining the period of oscillation from the timings of 20 oscillations with an electronically operated stopwatch of resolution 0.1 s (ignoring human reaction time in this example). In this case, all raw data and processed data (Table 5.3) are quoted to 3 sf.

Time for 20 oscillations				Period/s
Reading 1/s	Reading 2/s	Reading 3/s	Mean/s	
20.3	20.4	20.6	20.4	1.02

Table 5.3 Time for 20 oscillations

For a series of measurements, possibly for pendulums of different lengths, all the readings should be quoted to the resolution of the measuring instrument (and not necessarily the same number of sf). For example, Table 5.4 gives the stopwatch readings for a pendulum with a shorter length.

Time for 20 oscillations				Period/s
Reading 1/s	Reading 2/s	Reading 3/s	Mean/s	
8.7	8.4	8.6	8.6	0.43

Table 5.4 Time for 20 oscillations

The whole row of data is now quoted to 2 sf. This is different from the first set of readings. It would be incorrect to give this second set of readings to 3 sf, because this would imply a precision ± 0.01 s.

Accuracy and precision

Accuracy is how close a measured value is to an accepted literature or true value. The accuracy of your measurements depends on the instrument used, your skill as an experimenter and the techniques you are using.

Precision is a measure of the agreement among repeated measurements of the same quantity.

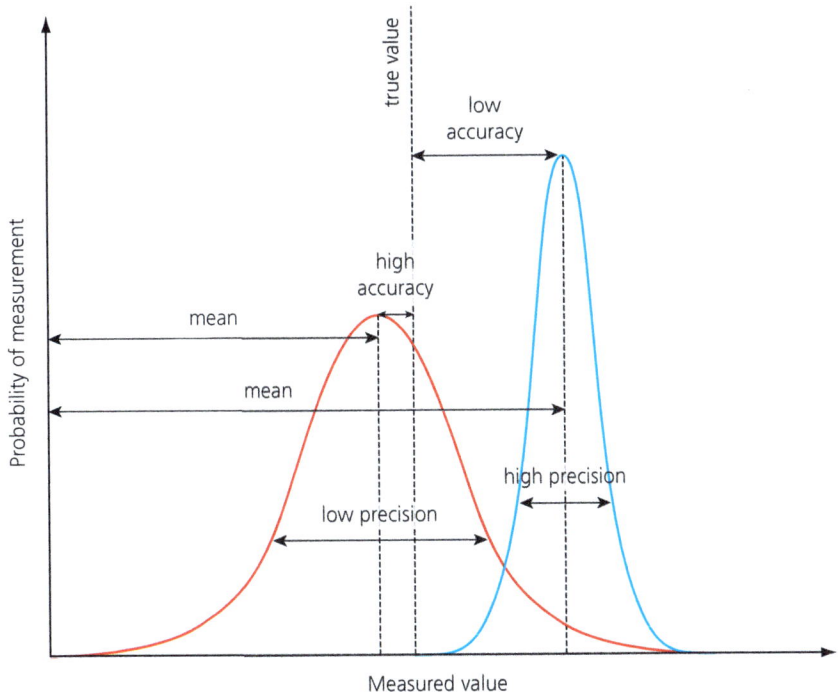

Figure 5.2 Accuracy versus precision: a set of readings may be very precise, but may not be accurate

> **Examiner guidance**
>
> There are two types of precision:
> - **Repeatability** is the precision when a single experimenter completes the investigation in a single session using the same equipment and instrumentation.
> - **Reproducibility** is the precision under any other set of conditions, including between different experimenters, or between different laboratory sessions for a single experimenter.

When an experimental measurement is repeated and recorded a number of times, the individual readings should all lie close to each other. Using an instrument whose measurements have less precision means that there will be a greater spread of readings, resulting in greater uncertainty in the measurement.

Errors

Any measured physical quantity, or calculated physical constant from experimental data, requires the following to be stated in order to be completely specified:

- a numerical value
- an associated unit
- the uncertainty (random error, taking into account any systematic error).

An example of a fully specified quantity might be:

acceleration = +11.4 km s^{-2} ± 0.2 km s^{-2} or (11.4 ± 0.2) km s^{-2}

> **Examiner guidance**
>
> Note that the two words 'uncertainty' and 'error' are often used interchangeably. 'Uncertainty' is best used to represent a range which contains the true value, whereas 'error' is best used to refer to systematic errors.

Uncertainties (or errors) can be specified as absolute, fractional or percentage uncertainties/errors, as shown in the following example.

> **Worked example**
>
> Time for object to fall = 14.32 s ± 0.02 s.
>
> The 0.02 s is referred to as the **absolute uncertainty** in the measurement. Note this has a unit.
>
> The fractional uncertainty = $\frac{0.02}{14.32}$ = 0.0014.
>
> The percentage uncertainty = fractional uncertainty × 100% = 0.14%. So the measurement can be expressed as: time to fall = 14.32 s ± 0.14%.

The uncertainty or error must be quoted to the same number of dp as the value, for example, 14.32 ± 0.02 and not 14.32 ± 0.023.

The quoted uncertainty in a quantity shows the number of sf its value should contain. For example, a measurement might be stated as (9.77 ± 0.01) m but *not* as (9.7743 ± 0.01) m, because the uncertainty of ±0.01 m clearly indicates that the 3rd sf in the value is uncertain and so it is meaningless to show the 4th and 5th.

■ Uncertainties for specific apparatus
■ Metre rule

The uncertainty in a length measurement with a metre rule can be taken in practice as ±1 mm. Remember that all length measurements using a metre rule actually involve two readings – one at each end – both of which have an uncertainty of half the smallest division (±0.5 mm).

> **Key definitions**
>
> **Repeatability** – Precision obtained when measurement results are produced in one laboratory, by a single experimenter, using the same equipment under the same conditions.
>
> **Reproducibility** – Precision obtained when measurement results are produced by different laboratories (or by different experimenters using different pieces of equipment).
>
> **Absolute uncertainty (absolute error)** – The uncertainty or error in a measurement that is expressed in physical units.

> **Expert tip**
>
> The uncertainty in a reading of a measurement is not totally due to the resolution of the scale. For example, a metal metre rule expands or contracts as its temperature increases or decreases. At only one temperature will the manufacturer claim the measurements to be accurate (within experimental error). At all other lower and higher temperatures there will be an uncertainty due to thermal expansion of the scale. Knowing the coefficient of thermal expansion would allow this uncertainty to be removed and precision improved.

> **Examiner guidance**
>
> When using scientific notation, quote the value and the error with the same exponent, for example, (6.6 ± 0.9) × 10^2 mm^2.

> **Examiner guidance**
>
> There are two different ways of indicating the random uncertainties in measurements:
>
> - Explicit – using plus or minus, ±, followed by a number.
> - Implicit – restricting the number of significant figures so that *only* the last digit is uncertain, which implies an uncertainty of ± 1 in the last decimal place.

In certain cases, the uncertainty may be greater than this because of some difficulty in measuring the required quantity, for example, due to parallax error or due to the short time available to record the reading. (When measuring the height of a ball's bounce the uncertainty could be ±5 cm.)

■ Standard masses

For 20 g, 50 g and 100 g standard masses the uncertainty can be taken as ±1 g, unless the manufacturer's value (often about 3%) is known.

■ Thermometers

For standard –10 °C to 110 °C mercury or alcohol thermometers the uncertainty can be taken as ±1 °C. The uncertainty of digital thermometers could be ±0.1 °C. However, the actual uncertainty in measured values may be greater than this due to thermal lag as an object is being heated or cooled (the object not being in thermal equilibrium with the thermometer).

■ Measuring cylinders/beakers/burettes/pipettes

For 25 cm^3 pipettes and 50 cm^3 burettes the resolution for a single measurement is ±0.05 cm^3. For measuring cylinders the resolution varies with volume and can be up to ±1 cm^3. In the case of measuring the volume using the line on a beaker, the uncertainty is likely to be much greater.

Beakers are for holding liquids, not for measuring volumes accurately.

■ Determining the period of oscillation of a pendulum or spring

To decrease uncertainty, often 20 oscillations are measured. The absolute error in the period (the time for a single oscillation) is then $\frac{1}{20}$ of the absolute error in the time for 20 oscillations.

> **Worked example**
>
> Time for 20 oscillations = 15.8 s ± 0.1 s
>
> Period = $\frac{15.8 \pm 0.1 \text{ s}}{20}$ = 0.790 s ± 0.005 s
>
> Note that the percentage uncertainty in the period is the same as that in the overall time:
>
> $\frac{0.1}{15.8} \times 100\% = 0.6\%$ (1 sf) and $\frac{0.005}{0.790} \times 100\% = 0.6\%$ (1 sf)

Expert tip

A fiducial marker (pointer) behind or below the object can help you judge when the oscillating object passes the centre/equilibrium position (Figure 5.3). The speed of the oscillator is greatest as it passes through the equilibrium position, so you should time from this point. If you try to time from the maximum displacement, the object will be moving very slowly and it will be difficult to judge when it has actually reached the point of maximum displacement and is about to reverse its direction.

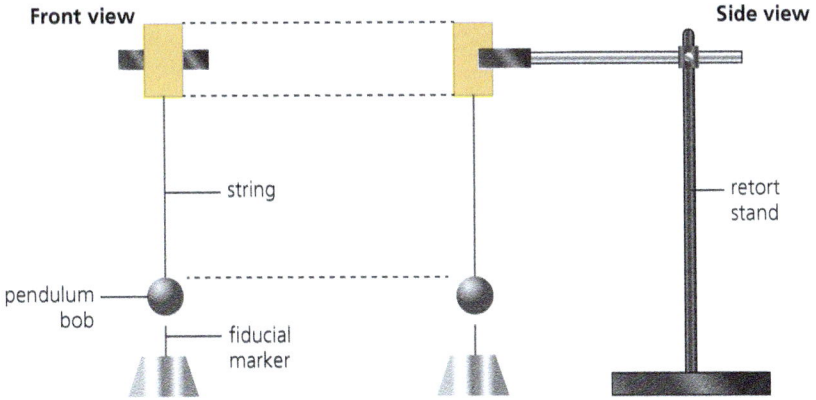

Figure 5.3 Use of a fiducial marker

■ Identifying and reducing random and systematic errors

Random errors are due to fluctuations in measurements, with an equal chance of being positive or negative. Taking the mean of repeat measurements therefore reduces the uncertainty due to these.

Systematic errors cannot be detected or reduced by recording more measurements and averaging. Figure 5.4 compares and contrasts random and systematic errors.

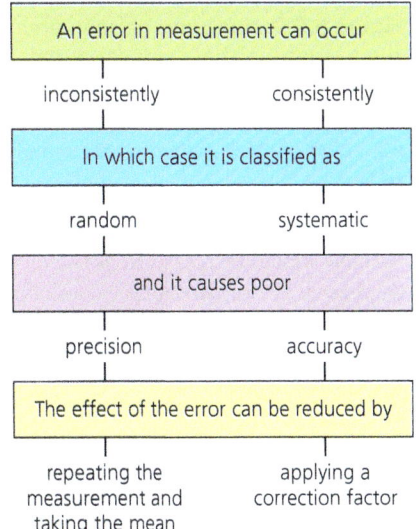

Figure 5.4 Comparing and contrasting random and systematic errors

> **Expert tip**
>
> - To reduce parallax errors when reading scales, any pointer should always be as close as possible to the scale and you should view the scale normally so your eye (line of sight) is perpendicular to the scale (see Chapter 2, Figure 2.2).
> - The scales on some meters are fitted with a small plane mirror to help view the scale normally. When the image of the pointer in the mirror is hidden behind the pointer, then the scale is being viewed correctly and the reading can be recorded (Figure 5.5).
>
>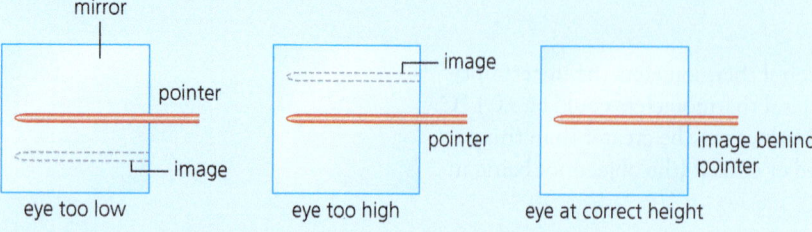
>
> **Figure 5.5** Use of a plane mirror to reduce parallax error

Measuring instruments are prone to variation. For example, both mechanical and electrical instruments will vary with the ambient temperature (and other factors), both analogue and digital instruments suffer from rounding errors, and low signal measurements are prone to the effects of noise (Figure 5.6).

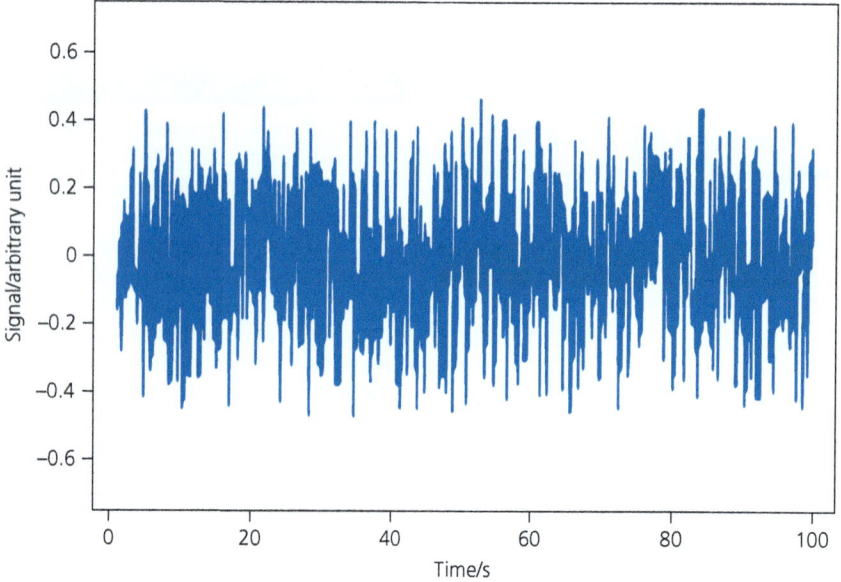

Figure 5.6 Background noise in an instrument showing the random fluctuations in the electrical signal

It is not always possible or appropriate to repeat a single reading and then average a number of measurements; for example, current and pd values to determine the resistance characteristics of an electrical component.

Particular values of one of the variables should be chosen, and then the other variable should be measured as the former variable is increased and then decreased. So, for particular values of pd, measure the current as the pd is increased and then decreased. The two current values for each pd value can then be averaged.

Another example is the recording of load and extension values to determine the spring constant of a metal spring. For particular values of the load added to the spring, measure the extension as the load is increased and then decreased. The values of extension should then be averaged.

When graphing experimental data (Figure 5.7), you can see immediately if you are dealing with random or systematic errors (provided you can compare with theoretical or expected results).

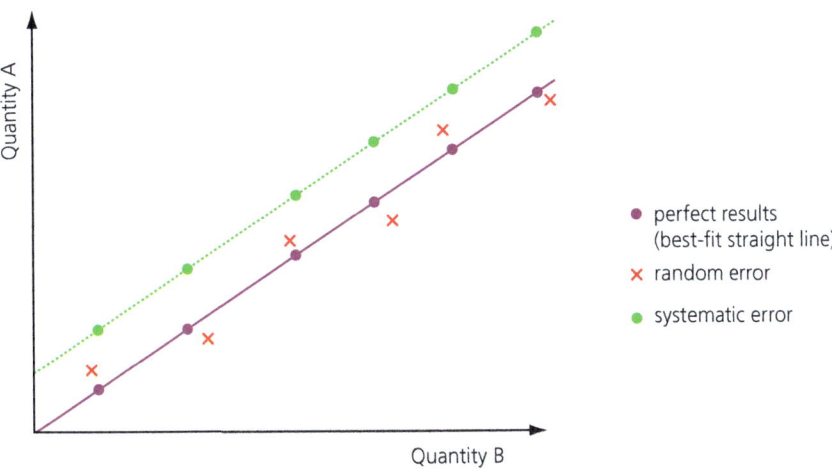

Figure 5.7 Effect of systematic errors on a linear relationship

> **Examiner guidance**
>
> Two approaches that can be used to reduce systematic errors are:
>
> - Compare the results to those of another experiment done using different apparatus and/or a different methodology.
> - Use the equipment to make measurements of known values. If there is good agreement there is greater confidence that the systematic error is insignificant and results are reliable.

Many physical quantities vary with temperature or pressure and if these are not constant during your investigation then the introduced error is likely to be partly random and partly systematic.

■ Independent and dependent errors

Consider a solid sphere of diameter 18.0 cm ± 0.2 cm. The calculated volume ($\frac{4}{3}\pi r^3$) is 3.1×10^3 cm³ ± 0.1 cm³. These uncertainties are dependent: if you overestimate the diameter, then you will calculate a larger value of the volume; if you measure a smaller volume, you will calculate a smaller diameter.

However, if you measure the mass of the solid sphere and determine it to be 13.0 g ± 0.1 g, then this is an independent error, because it comes from a different measurement, recorded with a balance.

If the uncertainties are due to errors in the measurement techniques, the error sizes of the variables of mass measurement and diameter measurement for the same sphere will be uncorrelated: a plot of mass versus diameter for solid spheres of different sizes will have no overall trend.

■ Estimating the uncertainty in the mean value of a measurement

Consider a quantity x that is measured several times during an investigation. A series of slightly different values is obtained: $x_1, x_2, x_3, \ldots x_n$, where n is the number of trials (which should usually be at least 3 and ideally 5). Unless there is a reason to suspect one of the results is anomalous, the best estimate of the true value of x is the **mean**, or average, of the readings:

$$\text{mean value } \bar{x} = \frac{x_1 + x_2 + \ldots + x_n}{n}$$

Any anomalous measurements are not included in this calculation (see 'Rejection of data' below).

A reasonable estimate of the uncertainty is half of the range of the data. That is, $\frac{x_{max} - x_{min}}{2}$, where x_{max} is the maximum and x_{min} is the minimum reading of x.

> **Key definition**
>
> **Mean** – An arithmetic average of a set of values. The mean of a set of repeated measurements (with random errors) will give a more accurate result.

5 Mathematical and measurement skills

> ■ **ACTIVITY**
>
> 5 The following results were obtained for the time taken for an object to roll down a slope. A manually operated electronic stopwatch was used for the time measurements. Determine the mean time and the associated uncertainty.
>
> 4.4 s, 4.8 s, 4.6 s, 5.2 s, 5.0 s.

■ Sources of error in addition to measurement uncertainty

Different investigations have different inherent sources of error depending on the instruments used and the procedure. These errors will make the uncertainty larger than expected from the resolution of the instruments. Table 5.5 shows some examples.

Type of experiment	Possible sources of error
Measuring the dimension of an object	The dimension of the object may not be uniform.
Experiment involving manual timing	Human reaction time will contribute towards random error and possibly systematic error (for example, by consistently overshooting).
Oscillating object	The presence of air resistance does not allow the object to oscillate freely.
Object moving along a surface	Friction with the surface will slow down the motion.
Heat experiments	There is transfer of thermal energy to or from the environment (surroundings).
Lens experiment	There is a range of distance in which the image appears sharp, hence it is difficult to determine the exact position where the image is the sharpest.
Experiments using optical pins	The holes made by the pins on the board are relatively large, making it difficult to construct the rays accurately.
Electricity experiment involving the use of resistance wire	The resistance wire is heated by the current so its resistance will change.
	Kinks in the wire will prevent the wire being straight along the ruler.
Electricity experiments in general	There is contact resistance at the points of connection.
	Batteries are not ideal – they have internal resistance.

Table 5.5 Some examples of inherent errors in physics investigations

■ Rejection of data

You may find during an investigation that one result in a set of measurements does not agree well with the other results – it appears to be an **outlier** (Figure 5.8). You must consider how large the difference between the suspect result and the other data is before discarding the result as anomalous. One simple approach is based on standard deviation.

Calculate the mean and **standard deviation** for your data. Any data value equal to or greater than two standard deviations from the mean value may be rejected with a high percentage of confidence.

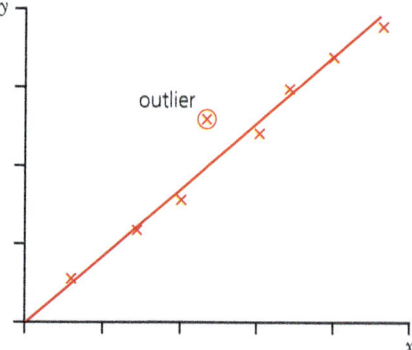

Figure 5.8 An outlier on a linear graph that could be an anomalous result

> ■ **ACTIVITY**
>
> 6 Consider the following set of measurements (ignoring units for simplicity). Apply the 'two standard deviation test' to decide whether the reading 0.1021 should be discarded.
>
> 0.1012, 0.1014, 0.1012, 0.1021, 0.1016

> **Key definitions**
>
> **Outlier** – Any value that is numerically distinct from most of the other data points.
>
> **Standard deviation** – A measure of the spread of a set of data from the mean.

Combining uncertainties in calculations

When processing data, the uncertainties in the values should be combined using the rules summarised in Table 5.6. This is called **propagation of errors**.

Combination	Operation to obtain uncertainty in result	Example
Adding or subtracting values $y = a \pm b$	Sum the absolute uncertainties of quantities. $\Delta y = \Delta a + \Delta b$	$y = (2.1 \pm 0.1)$ m $+ (3.4 \pm 0.4)$ m $- (0.82 \pm 0.01)$ m $= 4.7$ m ± 0.5 m
Multiplying or dividing values $y = \dfrac{ab}{c}$	Add the percentage (or fractional) uncertainties. $\dfrac{\Delta y}{y} = \dfrac{\Delta a}{a} + \dfrac{\Delta b}{b} + \dfrac{\Delta c}{c}$	$y = (2.1 \pm 0.1)$ m $\times (3.4 \pm 0.4)$ m $\div (0.82 \pm 0.01)$ m $\dfrac{\Delta y}{y} = \dfrac{0.1}{2.1} + \dfrac{0.4}{3.4} + \dfrac{00.1}{0.82} = 0.18$ therefore $y = 8.7$ m $\pm 18\%$ or (8.7 ± 1.6) m
Powers $y = a^n$	Multiply the percentage uncertainty by the power. $\dfrac{\Delta y}{y} = n\left\|\dfrac{\Delta a}{a}\right\|$	$y = (4.0 \pm 0.2)^2$ m $\times (2.2 \pm 0.3)^3$ m $\dfrac{\Delta y}{y} = 2\left(\dfrac{0.2}{4.0}\right) + 3\left(\dfrac{0.3}{2.2}\right) = 0.1 + 0.4 = 0.5$ therefore $y = 8$ m ± 4 m

Table 5.6 Summary of error propagation for random errors

ACTIVITY

7 Given $a = (3.2 \pm 0.2)$ and $b = (2.3 \pm 0.1)$, calculate $a + b$, $a - b$, $a \times b$ and $\dfrac{b}{a}$ and deduce the overall uncertainty in the result of each operation.

8 The diameter of a circle is measured as (4.6 ± 0.2) mm. Calculate the radius and the circumference and associated uncertainties in the quantities calculated.

Error propagation for logarithmic and trigonometric functions

For trigonometric and logarithmic functions, you need to determine the following:

|max value − mean value|

|min value − mean value|

Whichever is larger is the ± reported uncertainty. This can be time consuming as it does not involve percentage uncertainties in the final uncertainty determination.

Expert tip

It is a good idea to propagate your uncertainties as you work on your individual investigation. This enables you to see which of the various sources of uncertainty is having the greatest effect, and so gives you a chance to modify your methodology.

Key definition

Propagation of errors – Calculation of the overall uncertainty when processing data containing random errors through a sequence of mathematical operations.

Worked example

If $k = 4.78 \pm 0.35$ cm^{-1} and $x = 23.5 \pm 0.1$ cm, determine $\sin(kx)$ and its uncertainty.

The maximum value of kx is $5.13 \times 23.6 = 121.068$.
  $\sin(121.068°) = 0.85656$ (minimum value)

The minimum possible value of kx is $4.43 \times 23.4 = 103.662$.
  $\sin(103.662°) = 0.97171$ (maximum value)

Using the stated values of k and x, $kx = 112.33$.
  $\sin(112.33°) = 0.92501$
  |max value − mean value| = |0.97171 − 0.92501| = 0.04670
  |min value − mean value| = |0.85656 − 0.92501| = 0.06845

The final answer is therefore 0.925 ± 0.068, which is best reported as 0.93 ± 0.07.

Graphs

Graphs show relationships between two variables and are used for deriving other data (for example, from the gradient, area or intercept). The independent variable is plotted along the x-axis and the dependent variable along the y-axis. Where appropriate, **error bars** should be included in a graph of experimental data (Figure 5.9). These show the uncertainty in the x and y values for each plotted data point.

> **Key definition**
>
> **Error bars** – Graphical representation used on graphs to display the uncertainty (error) in a measurement.

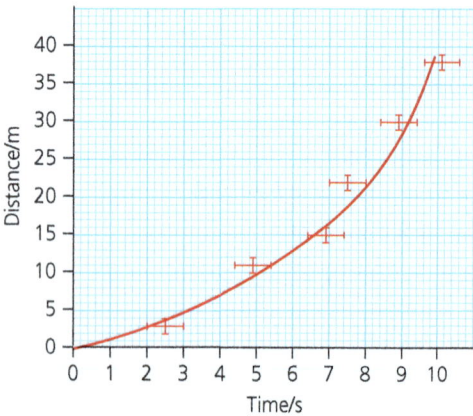

Figure 5.9 Error bars in a graph of distance versus time for an accelerating object

> **Worked example**
>
> Determine the uncertainty in ln F that should be used to plot an error bar if $F = 8.2\ N \pm 0.2\ N$.
>
> $\ln(8.2) - \ln(8.0) = 2.104 - 2.079 = 0.025$
>
> $\ln(8.4) - \ln(8.2) = 2.128 - 2.104 = 0.024$
>
> The value $\ln(8.2) = 2.10$ should be plotted with an error bar between 2.08 and 2.13.

> **Examiner guidance**
>
> Current–voltage characteristics and graphs investigating the relationship between the extension of a spring and the force applied to the end (Hooke's law) are often plotted with the dependent variable (current and extension in these cases) on the x-axis. This makes the gradient (the constant of proportionality) equal to the properties of electrical resistance and the spring constant.

■ Determining the gradient

The gradient of any graph represents the change in y for a corresponding change in x:

$$\text{gradient} = \frac{\Delta y}{\Delta y}$$

It is the rate of change and is determined graphically for a straight line graph by drawing a large triangle using the **line of best fit** as the hypotenuse of the triangle, as shown in Figure 5.10. In this example, the gradient is equal to the rate of change of velocity with time, $\frac{\Delta v}{\Delta t}$, or acceleration.

> **Key definition**
>
> **Line of best fit** – A graph line drawn to pass through the plotted points, so that most lie on the line or roughly evenly distributed on either side of the line.

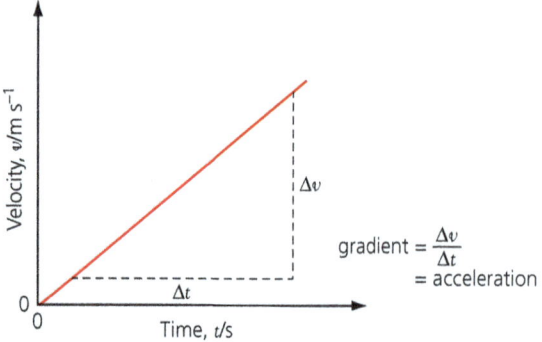

Figure 5.10 Determining the gradient of a velocity–time graph for an object with uniform acceleration

To estimate the gradient at a particular data point on a curved graph, draw a tangent to the curve at that point. Then draw a large triangle using the tangent as the hypotenuse, as shown in Figure 5.11, and use this to determine the gradient.

■ Error bars and uncertainties in the gradient

To determine the uncertainty in the gradient of a linear graph with error bars, two gradients should be drawn on the line graph as shown in Figure 5.12. One gradient line (blue) is the steepest and the other (red) is the shallowest gradient line that can be drawn through the error bars. The line of best fit would be drawn between these two.

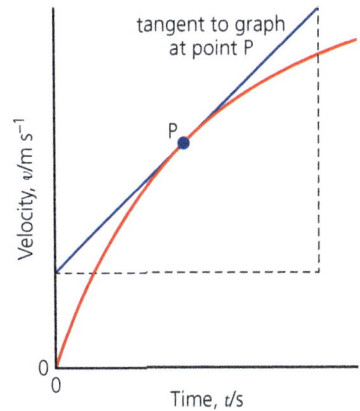

Figure 5.11 Determining the gradient of a velocity–time graph for an object with changing acceleration; the gradient of the tangent at point P gives the acceleration at that time

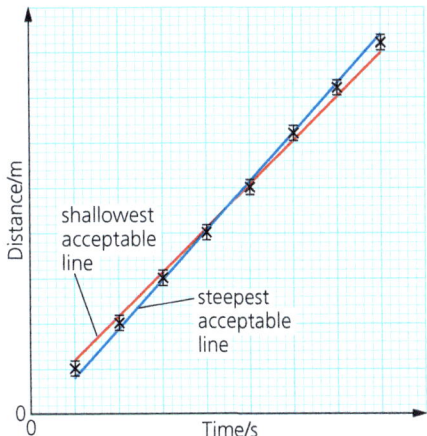

Figure 5.12 Shallowest gradient and steepest gradient for a distance–time graph

The gradient of each line can be determined and then substituted into the following equation:

$$\% \text{ uncertainty in gradient} = \frac{\text{highest (or lowest) gradient} - \text{best fit gradient}}{\text{best fit gradient}} \times 100$$

Whether the highest or lowest gradient is used depends on which gives the greater percentage uncertainty.

■ ACTIVITY

9 Estimate the uncertainty in the value of the gradient of the data shown in Figure 5.13 and hence the uncertainty in the value for the capacitance.

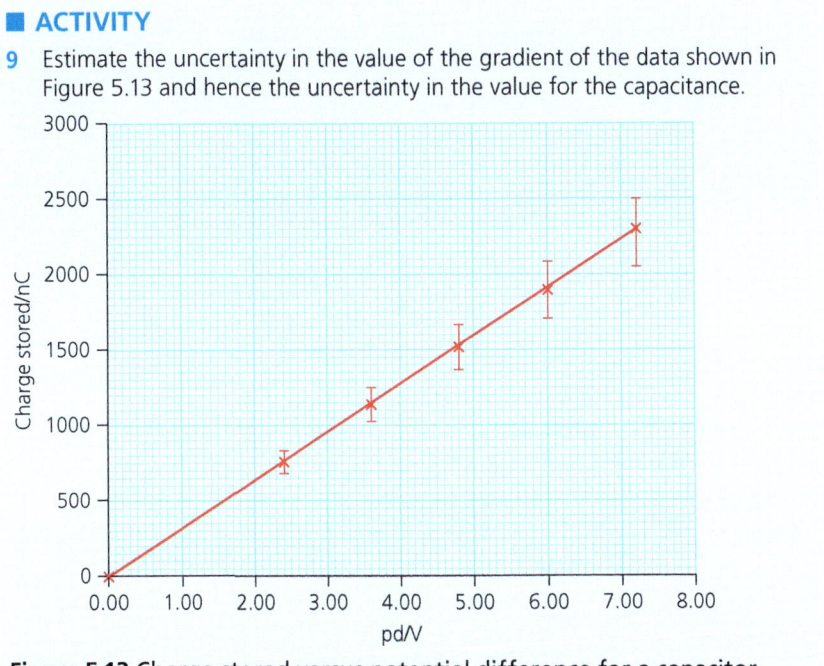

Figure 5.13 Charge stored versus potential difference for a capacitor

An Excel function that can be used to estimate the uncertainty in a gradient is LINEST. Consider the results in Table 5.7 entered into a spreadsheet. Current values are entered in cells A2–A9 and voltage values in B2–B9.

Current/mA	Voltage/V
0.000	0.00
0.069	0.50
0.140	1.00
0.242	1.50
0.309	2.00
0.359	2.50
0.460	3.00
0.516	3.50

Table 5.7 Current and voltage readings for graphing by Excel

The gradient of the graph of voltage against current is the resistance. The LINEST function determines the gradient of the line of best fit and estimates the uncertainty in this gradient.

Select four empty cells below the table of results and highlight them and type in the formula shown in Figure 5.14.

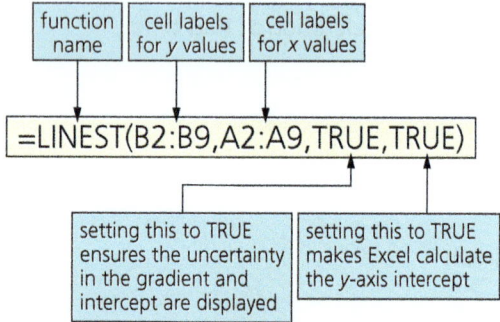

Figure 5.14 LINEST function in Excel

Then hold down CTRL and SHIFT and press ENTER on a PC (or press CMD + SHIFT on an i-Mac). Four values will be displayed in the four highlighted cells (Figure 5.15). Taking into account significant figures, the resistance is 6.6 ± 0.2 Ω.

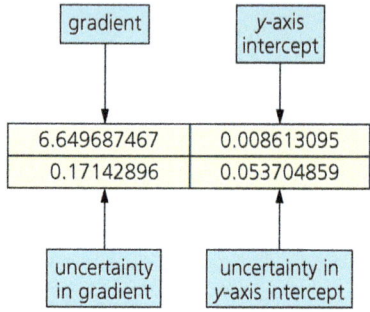

Figure 5.15 Output of LINEST function

■ Types of graph

The most useful form of a graph is a straight line, which is described by the general equation $y = mx + c$, where x and y are variables and m and c are constants. Figure 5.16 shows that when $x = 0$ the intercept on the y-axis is c; when $y = 0$ the intercept on the x-axis is $-\frac{c}{m}$. The gradient of the line (the change in y with x, $\frac{\Delta y}{\Delta x}$), is m.

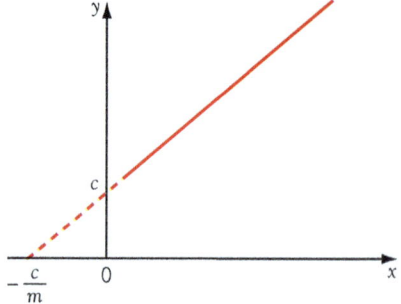

Figure 5.16 Graph of $y = mx + c$

The general equation $y = mx^2 + c$ defines a quadratic function (Figure 5.17). If $c = 0$ the graph passes through the origin. An example of this would be the variation of kinetic energy with velocity of a body in simple harmonic motion.

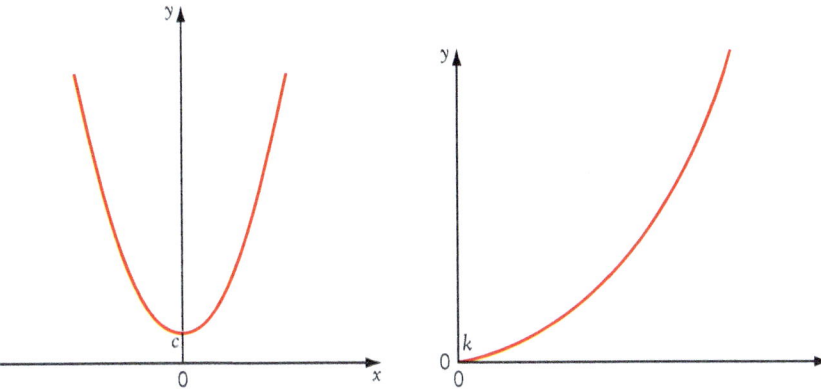

Figure 5.17 Graph of $y = mx^2 + c$

Figure 5.18 Graph of $y = ke^x$

$y = ke^x$, where k is a constant, is an exponential function (Figure 5.18). It describes an exponential increase in y with respect to x. An example of this would be the increase in the pressure of air with vertical depth (in the presence of a gravitational field).

$y = ke^{-x}$ is another exponential relationship, describing a decrease of y with respect to x (Figure 5.19). Equations of this type arise in radioactive decay and the discharge of a capacitor.

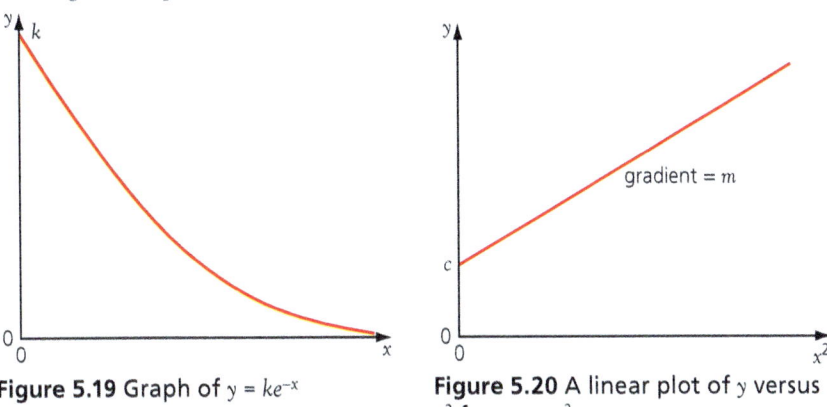

Figure 5.19 Graph of $y = ke^{-x}$

Figure 5.20 A linear plot of y versus x^2 for $y = mx^2 + c$

It is often useful to plot the results of an experiment in the form of a straight line. Quadratic and exponential equations can be mathematically transformed to give a linear relationship between two variables.

- For $y = mx^2 + c$, plot y against x^2 (Figure 5.20) to give a linear graph.
- For $y = k\,e^x$, plot $\ln(y)$ against x.
- For $y = k\,e^{-x}$, plot $\ln(y)$ against x.

This is known as linearization. It allows physical interpretation of information from the gradient and intercept (without the need for calculus).

For $y = k\,e^{cx}$ (where k and c are both constants), taking natural logarithms of both sides of the equation gives $\ln y = \ln k + cx$, and plotting $\ln y$ against x gives a straight line with gradient c and y-intercept of $\ln k$ (Figure 5.21).

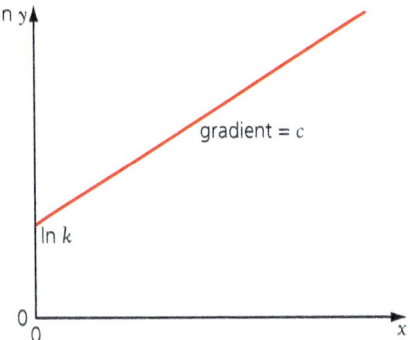

Figure 5.21 A linear plot of $\ln y$ versus x for $y = ke^{cx}$

Log graphs can also be used for power-law equations. Consider $y = kx^2$; taking logarithms (to base 10) of both sides gives $\log y = \log k + 2 \log x$. Plotting $\log y$ against $\log x$ (Figure 5.22) generates a straight line of gradient 2, with an intercept on the y-axis of $\log k$.

> **Examiner guidance**
>
> Plotting a log–log graph is a useful approach when deriving an unknown equation from a set of experimental results. Consider $y = ax^b$, where a and b are unknown constants. Taking logarithms (to base 10) of both sides gives $\log y = \log a + b \log x$. If you plot $\log y$ against $\log x$ you will get a straight line of gradient b and intercept on the y-axis of $\log a$ (Figure 5.23). The values of a and b from the graph allow the form of the equation to be determined.

> **Expert tip**
>
> Linearizing a graph means adjusting the variables mathematically so that the relationship between them generates a straight-line graph. It does *not* mean fitting a straight line through data points that form a curve.

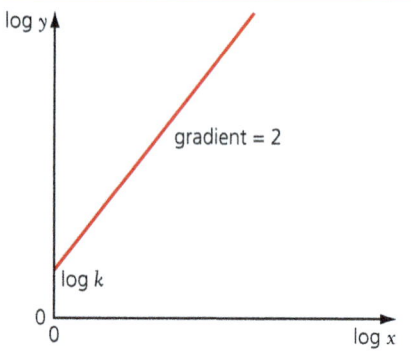

Figure 5.22 A linear plot of $\log y$ versus $\log x$ for $y = kx^2$

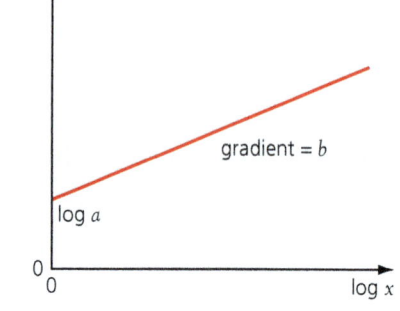

Figure 5.23 A plot of $\log y$ against $\log x$ for $y = ax^b$

Trigonometric graphs

Tangent (tan), sine (sin) and cosine (cos) are trigonometric functions. The sine and cosine functions (Figure 5.24) and the squares of these functions (Figure 5.25) are periodic, meaning they repeat every 360° (or 2π radians).

Graphs of $y = \sin x$ and $y = \cos x$ show how these functions change as the angle x changes. The two plots have the same form but are phase shifted relative to one another along the x-axis by 90° ($\frac{\pi}{2}$ radians). Both functions have values that range between +1 and −1.

> **RESOURCES**
>
> https://www.geogebra.org/download

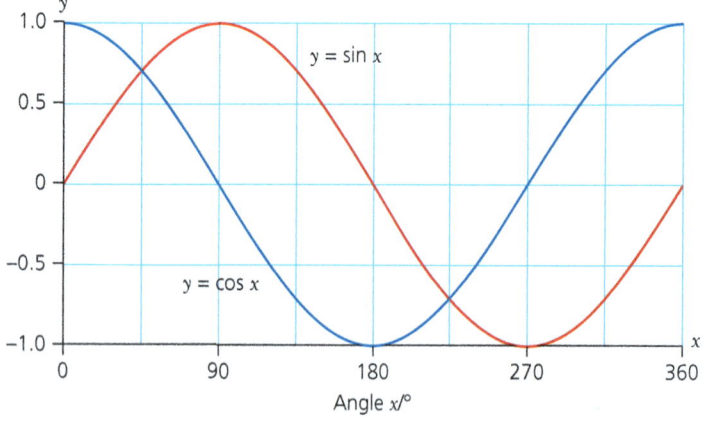

Figure 5.24 Graphs of $y = \sin x$ and $y = \cos x$

> **Expert tip**
>
> Degrees and radians are both used to measure angles. The radian is often a useful measure, especially when working with waves and oscillations.

> **Expert tip**
>
> A general form of the sine function is: $y = A \sin ax$, where A and a are constants.

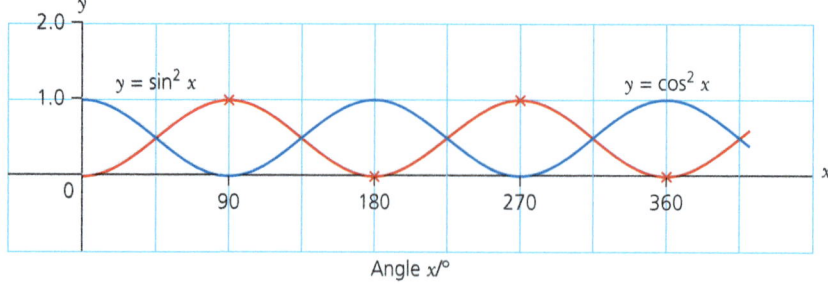

Figure 5.25 Graphs of $y = \sin^2 x$ and $y = \cos^2 x$

Whenever a physical phenomenon repeats itself in time, expect to see sin or cos functions in its mathematical description (model). Examples include waves, the behaviour of elementary particles (both classical and quantum), planetary motion, signal processing, and optics.

> **Examiner guidance**
>
> When using small angles in many areas of physics, including mechanics, electromagnetics and optics, you can use the following small-angle approximations:
>
> $\sin \theta \approx \theta$ (in radians)
>
> $\tan \theta \approx \theta$ (in radians)
>
> $\cos \theta \approx 1$

Other graphs

It may not always be possible or appropriate to plot a straight line graph. In addition, the relationship you are investigating for your individual investigation may not be described or fitted by any of the functions (forms) previously discussed.

For example, you may be investigating the variation in the output power of a battery as the external resistance varies. This relationship is shown in Figure 5.26. The power output increases rapidly to a maximum and then decreases less rapidly. The important region is near the maximum. Extra data points should be recorded and plotted near the maximum to enable a more precise value of the maximum to be estimated.

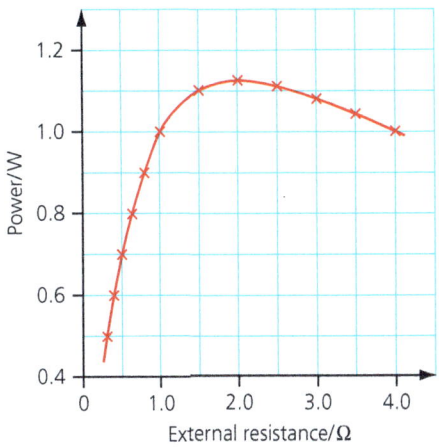

Figure 5.26 Output power of a battery as a function of external resistance

Beware of assuming that all relationships you investigate can have their raw data processed to generate a linear relationship. Figure 5.27a shows a graph of processed data of the rate of flow, Q, of a liquid along a tube as a function of the pressure difference, Δp, between the ends of the tube. The theory of streamline or laminar flow from fluid dynamics suggests that the relationship between Q and Δp should be of the form $Q = K \Delta p$, where K is a constant. You may consider that the data points lie close enough to a line of best fit to support the theoretical model. However, at high flow rates the flow of liquid is turbulent, and the graph of Q against Δp will be hyperbolic, approaching a plateau. The inclusion of error bars for the pressure readings in Figure 5.27b shows that the uncertainty in the individual readings is low, supporting the suggestion of a transition from laminar to turbulent flow.

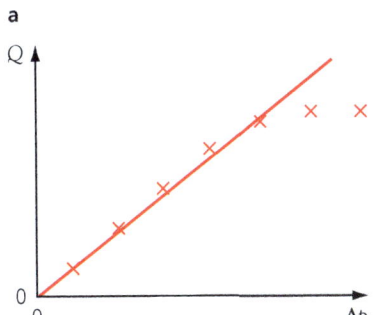
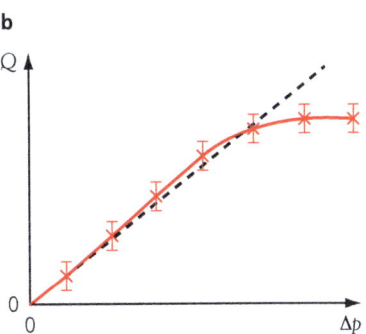

Figure 5.27 a Incorrect linear relationship; **b** correct relationship between Q (flow rate along tube) and Δp (pressure differential between the ends of the pipe)

6 Information communication technology (ICT)

Spreadsheets

■ Creating graphs with Excel

Select the block of cells containing the numbers to be plotted. The 'x-axis' data column should be to the left of the 'y-axis' data column. Click on the Insert tab and choose a scatter graph. Graphs generated can be easily copied or pasted into a Word document, where they can be edited.

■ ACTIVITY

1 Use Excel to generate a scatter plot for the experimental data in Table 6.1. Use the Internet or Excel Help function to find out how to add a linear trend line. Display the equation of the line and the correlation value.

Time/s	Distance/m
0	0.44
1	1.12
2	2.25
3	3.65
4	4.87
5	5.62
6	6.69
7	7.87
8	8.12
9	9.33
10	10.87

Table 6.1 Experimental data

Expert tip

A **correlation coefficient** is a number between −1 and +1 and shows the extent to which the two variables are linearly related. A perfect correlation with a positive gradient has a +1 correlation, while a negative gradient with a perfect correlation has a value of −1. A zero correlation means that there is no linear relationship. Most of your experimental data should have high correlations, such as 0.90 or 0.95, but this does not mean your linear fit is based on high quality data. You must consider systematic and random errors in the data.

■ Analysing data with Excel

If you want to write a formula for the expression $\frac{x^2 y}{z + w}$, where x is in cell A1, y is in cell A2, z is in cell B3 and w is in cell B5, you can write the formula in cell C2 as =A1^2*A2/(B3+B5).

If you have typed a formula in cell B1 and you want the same formula in cells B2 through B10, position the cursor in the lower right-hand corner of selected cell B1 until the cursor becomes a black cross. Then click and drag from cells B2 through B10. Note that any cells written in the formula of B1 will be shifted when the formula is copied (unless those cells are written with $ symbols, such as A2). For example, if B1 contains =2*A1 and you copy B1 to B2, then B2 will contain =2*A2.

■ Built-in functions in Excel

Excel has a number of built-in functions (Table 6.2) that can be used to process data and carry out statistical analysis. The function wizard *fx* on the toolbar will bring up a dialogue box where all the built-in functions are listed. Note that for the trigonometric functions the angles are, by default, in radians. You can use the DEGREES (angle) function to convert radians into degrees.

Key definition

Correlation coefficient – statistical measure that indicates the degree of relationship between two variables.

RESOURCES

Excel resources:
- http://science.clemson.edu/physics/labs/tutorials/excel/symbols.html
- http://science.clemson.edu/physics/labs/tutorials/excel/advanced.html

Excel formula	Description of the formula
=SUM(A2:A5)	Find the sum of values in the range of cells A2 to A5.
=COUNT(A2:A5)	Count the number of numbers in the range of cells A2 to A5.
=COUNTIF(A1:A10,100)	Count cells equal to 100.
=COUNTIF(A1:A10,">30")	Count cells greater than 30.
=AVERAGE(A2:A5)	Find the mean (average) of the numbers in the range of cells A2 to A5.
=ABS(A2)	Find the absolute value of the number in cell A2.
=SQRT(A2)	Find the square root of the number in cell A2.
=LN(A1)	Find the natural logarithm of the number in cell A1.
=LOG(A1)	Find the logarithm (to base 10) of the number in cell A1.
=EXP(A1)	Returns e raised to the power of the number in cell A1.
=SIN(A1)	Finds the sine of the number in cell A1.
=TAN(A1)	Finds the tangent of the number in cell A1.
=COS(A1)	Finds the cosine of the number in cell A1.
=PI()	Returns the value of π: 3.141 592 654.
=POWER(x,n)	Returns the number x raised to the power n.

Table 6.2 Built-in Excel functions

Simulations with Excel

Models or **simulations** of physical processes are classified as either deterministic or random.

An example of a deterministic process is planetary motion and its description using Newton's equations of motion: at any time the position and momentum of a planet in the system at time t can be used to predict the position and momentum of the planet at a later time t'.

Many physical systems are inherently disordered and cannot be analysed deterministically, such as electric breakdown in dielectrics and galaxy formation. A good example of a physics theory based on randomness at the atomic level is radioactive decay. Individual decays cannot be predicted but a large number of them can be described statistically.

To model randomness in Excel (or in a programming language such as Python or Visual Basic), a random number generator is needed, to simulate the rolling of dice or tossing a coin. There is a wide range of calculations in computational physics that use a random number generator and are known as Monte Carlo simulations. This approach also allows the inclusion of temperature in modelling.

The radioactive decay process can be simulated for one nucleus by generating a random number in the range 0 to 1 and comparing this with the probability that the nucleus decays within a time interval. If the random number is less than this probability then the radioactive decay occurs; if not the nucleus does not decay.

Excel provides two functions for generating random numbers:

- RAND() generates a random decimal number between 0 and 1.
- RANDBETWEEN(a, b) generates a random integer between a and b.

These functions are volatile, meaning that every time there is a change to the worksheet their value is re-calculated and a different random number is generated.

RANDBETWEEN only generates integer values. If you want a random number which could be any decimal number between a and b, then use the following formula instead:

=a+(b−a)*RAND()

> **Key definition**
>
> **Simulation** – A representation (model) of a process or a system usually involving ICT that imitates a real or an idealized situation.

> **RESOURCES**
>
> Excel and physics resources:
> - http://mmsphyschem.com/excelPhys1.htm
> - http://mmsphyschem.com/excelPhys2.htm
> - http://academic.pgcc.edu/~ssinex/UCCEN_F05/

6 Information communication technology (ICT)

> ■ **ACTIVITY**
>
> 2 The Analysis ToolPak is used for statistical analyses or engineering analyses. On a PC, open a new Excel spreadsheet and click File, Options and then Add-Ins. In the Manage box, select Excel Add-ins and click Go. Select 'Analysis ToolPak' and click OK.
>
> On a Mac in the file menu go to Tools and then Excel Add-ins. Open a blank spreadsheet in Excel and fill column A with 10 000 random numbers in the interval −1 to 1. Use the Data/Data Analysis/ Random Number Generation dialogue box with:
>
> - number of variables = 1
> - number of random numbers = 10 000
> - distribution = uniform
> - parameters = between −1 and 1
> - random number seed = any integer value
> - output range = select cell for first random number
>
> Use the AVERAGE and STDEV functions to find the average and standard deviation of this sample.

Expert tip

The distribution that was generated is uniform in the range −1 to 1. This means that there should be equal numbers of numbers in equal-sized intervals.

A mathematical description of radioactive decay (for a system of $N(0)$ particles at time $t0$) is given by:

$$P(t) = \frac{N(t)}{N(0)} = e^{-\lambda t}$$

The probability that the system has not decayed after a time $t + \delta t$, $P(t + \delta t)$, is equal to the product of the probability that it has survived a time t, $P(t)$, and the probability that it does not decay in the time interval δt:

$$P(t + \delta t) = P(t)(1 - \lambda \delta t)$$
$$= P(t) - \lambda P(t)\delta t$$
$$\frac{P(t + \delta t) - P(t)}{\delta t} = -\lambda P(t)$$

In the limit $\delta t \to 0$, $\frac{dP}{dt} = -\lambda P(t)$ which when integrated gives $P(t) = e^{-\lambda t}$.

You can simulate the radioactive decay process (Figure 6.1) by determining whether a radioactive nucleus has decayed during a small interval of time δt.

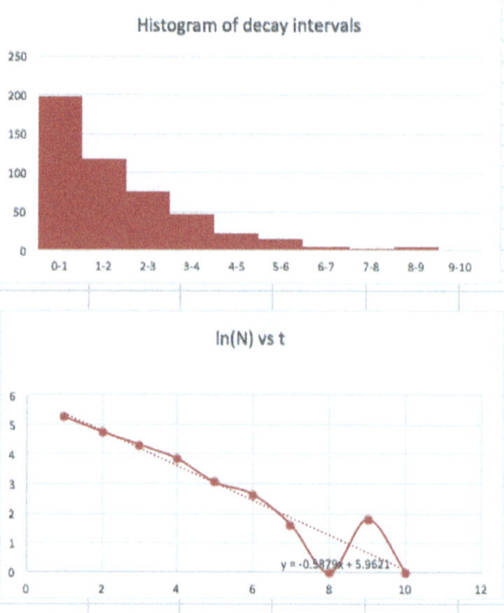

Figure 6.1 Excel simulation of radioactive decay

If it does not decay it passes on to the next time interval and this is repeated until the nucleus has decayed. The spreadsheet then sums all the time intervals to find its lifetime, which is then repeated for many nuclei.

Successive rows in the spreadsheet represent successive time intervals of length δt. In each cell in one column you can enter a uniform random number in the range 0–1. If this value is greater than or equal to $\lambda \delta t$ the nucleus does not decay and survives to the next time interval. The process is repeated in successive spreadsheet cells until the random number is less than the value of $\lambda \delta t$ and the nucleus decays. Then the program moves on to examine another nucleus. The sum of time intervals each nucleus passes through before decaying gives its total lifetime. This whole process is repeated for a large number of nuclei and histograms and plots generated.

■ ACTIVITY

3   Open a new workbook in Excel and use some cells to enter values for δt and λ, and their product $\lambda \delta t$. In column A enter the numbers 1, 2, … 10 000. These represent 10 000 time intervals. In the adjacent column put 10 000 random numbers from a uniform distribution between 0 and 1. (Use the random number generator in the Data Analysis Toolkit or the =RAND() function.) In the next two columns enter formulae to work out whether the current nucleus decays in each interval and when to move on to the next nucleus.

The 'Nucleus' column will show a series of 1s, followed by 2s, 3s, and so on. These correspond to nuclei 1, 2, 3, … and the number of successive cells containing the same nucleus number gives the total number of time intervals through which the nucleus has passed.

The maximum number of nuclei whose decay may be simulated in this activity (with 10 000 random numbers) depends on the choice of values for δt and λ. For $\delta t = 0.1$ and $\lambda = 0.5$, there are ~500 nuclei.

Row/Col	A	B	C	D
1	δt [s]	0.1		
2	λ [s^{-1}]	0.5		
3	$\lambda \delta t$	0.05		
4				
5	Time interval	Random number	Decay?	Nucleus
6	1	0.548412494	=IF($B6<$B$3,"Y","N")	1
7	2	0.134629132	=IF($B7<$B$3,"Y","N")	=IF(C6="N",D6,D6+1)
8	3	0.317186993	=IF($B8<$B$3,"Y","N")	=IF(C7="N",D7,D7+1)
9	4	0.014117541	=IF($B9<$B$3,"Y","N")	=IF(C8="N",D8,D8+1)

Table 6.3

Row/Col	F	G	H
5	Nucleus	Intervals	Lifetime [s]
6	1	=COUNTIF(D6:D10005,F6)	=(G6-0.5)*B1
7	2	=COUNTIF(D7:D10006,F7)	=(G7-0.5)*B1
8	3	=COUNTIF(D8:D10007,F8)	=(G8-0.5)*B1
9	4	=COUNTIF(D9:D10008,F9)	=(G9-0.5)*B1

Table 6.4

The formulae in the 'Intervals' column count the number of time intervals before a specific nucleus has decayed. The formulae in the 'Lifetime (s)' column calculate the lifetime of a specific nucleus. (Activity continues on next page.)

Row/Col	J	K	L	M
5	Lifetime	Bin	Frequency	Ln(N)
6	0–1	1	{=FREQUENCY(H6:H488,K6:K15)}	=LN(6)
7	1–2	2	{=FREQUENCY(H6:H488,K6:K15)}	=LN(7)
8	2–3	3	{=FREQUENCY(H6:H488,K6:K15)}	=LN(8)

> **Expert tip**
>
> A plot of ln N (y axis) versus t (x axis) generates a straight line (Figure 6.1) which shows the rate of decay remains constant. It has a slope of $-\lambda$.

9	3-4	4	{=FREQUENCY(H6:H488,K6:K15)}	=LN(9)
10	4-5	5	{=FREQUENCY(H6:H488,K6:K15)}	=LN(10)
11	5-6	6	{=FREQUENCY(H6:H488,K6:K15)}	=LN(11)
12	6-7	7	{=FREQUENCY(H6:H488,K6:K15)}	=LN(12)

Table 6.5

The Excel FREQUENCY function returns a frequency distribution which is a summary table that shows the frequency of each value in a range. This function is =FREQUENCY (data_array, bins_array), where:

- data array is the worksheet range that holds the values for which you want to obtain frequencies

- bins_array is the worksheet range ('bins') for grouping values.

For example, the data array could be test scores, e.g. 75, 89, 55, 65, etc and the bins, e.g. 50-60, 60-70 etc.

The braces ({}) in the formulas of the 'Frequency' column indicate that the formula is a CSE function – once the bins (intervals) are created the formula is entered into cell L6 (do not press Enter). Then highlight the column from L6 to L12 and press CRTL-SHIFT-ENTER to copy the formula down the column.

Python programming

Learning some basic coding skills is a very worthwhile endeavour. More than 50% of jobs require some degree of technology skills and this percentage looks set to increase in the coming years, particularly in the STEM (science, technology, engineering, mathematics) fields. This is a good reason to gain some experience with basic coding skills while studying IB Diploma group 4 subjects.

Many topics in physics lend themselves to coding-type activities, including modelling and graphing kinematics, creating visualizations of fields (electric, magnetic or gravitational) and determining integrals or derivatives. Indeed, modelling a 'theoretical curve' for comparison with experimentally derived data could serve as a useful adjunct for many different internal assessment (IA) investigations.

One language that lends itself to this kind of visual modelling is Python or, more specifically, VPython, which is the Python programming language plus a 3D graphics module called Visual. There are a number of online IDEs (integrated development environments) where students can code using Python and/or VPython without needing to download any software. These include Cloud9 (c9.io), GlowScript (glowscript.org) and Trinket (trinket.io).

As an indication of the power of VPython, the short block of code in Figure 6.2 was used to create a graph and table of data illustrating Coulomb's law. The output of the code is shown in Figure 6.3. Note that the data in the window at the bottom of the screen may be copied and pasted into a spreadsheet for further analysis.

```
1   GlowScript 2.6 VPython
2
3   g = graph(title = "Coulomb's Law",xtitle="Distance (\u212B)",ytitle="Force (N)") #Note that \u212B is the unicode for Angstrom symbol
4   f1 = gdots(color=color.green) #sets the type of curve to be plotted to gdots
5   k = 8.99E9 #Coulomb's constant
6   e = 1.602E-19 #Charge on electron
7   Q1 = 1 #Charge on first particle in multiples of electron charge
8   Q2 = -2 #Charge on second particle in multiples of electron charge
9   d = 0.5 #Starting distance between the charges in Angstroms (10^-10 m)
10  delta_d = 0.1 #Distance increment
11  max_d = 6 #Maximum distance in Angstroms
12
13  print('Distance (\u212B)', 'Attractive Force (N)') #Change label to 'Repulsive Force' if Q1 and Q2 have the same sign
14  while d < max_d:
15      rate(10) #Sets the number of calculations per second
16      numerator = k * (abs(Q1) * e) * (abs(Q2) * e) #calculates the numerator
17      denominator = (d * 1e-10)**2 #calculates the denominator
18      F = numerator/denominator #calculates the force
19      print(d,'         ',F) #prints data table headings
20      f1.plot(d,F) #plots distance and force to graph
21      d += delta_d #increments the distance
```

Figure 6.2 VPython code for Coulomb's law simulation

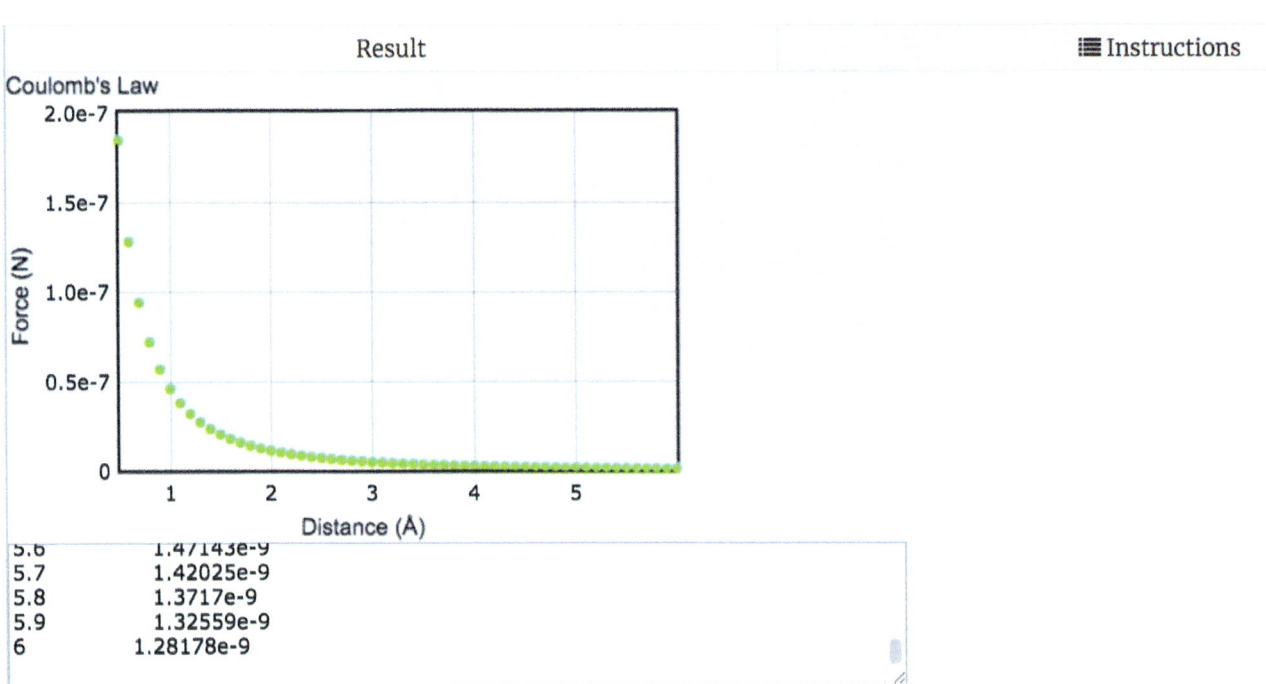

Figure 6.3 Output for Coulomb's law simulation

> **Expert tip**
>
> There are some excellent introductory videos on how to use VPython and GlowScript on YouTube and at **www.glowscript.org**.

The VPython code in Figure 6.4 was used to model a ball on a spring undergoing simple harmonic motion (SHM).

```
1  GlowScript 2.6 VPython
2
3  g1 = graph(title = "Displacement-time graph for a ball on a spring",xtitle="time (s)",ytitle="Displacement (cm)") #Sets up graph
4  f1 = gcurve(color = color.red) #Sets type of graph and colour of curve
5  #f2 = gcurve(color = color.blue)
6
7  L0 = 0.1 #the natural length of the spring
8  k = 10 #Set the force constant of the spring
9  t = 0 #Starting time
10 dt = 0.01 #Set the time increment
11 holder = holder= box(pos=vec(-.1,.1,0), size=vec(.1,.005,.1))
12 ball = sphere(pos = vector(-0.1,-L0+0.1,0), color = color.yellow, radius = 0.02)
13 spring=helix(pos=holder.pos, axis=ball.pos-holder.pos, radius=.02, coils=10)
14 ball.m = 0.1 #Sets the mass of the ball
15 g = vec(0,-9.82,0) #gravitational field
16 ball.v = vector(0,0,0) #Sets the initial velocity of the ball
17
18 while t < 10:
19     rate(100) #Sets the calculation rate to 100 calculations per second (i.e. real time data)
20     L = ball.pos - holder.pos #L is the length of the spring from the holder to the ball
21     F = -k *(mag(L)-L0)*norm(L) #Calculates the restoring force on the ball; mag(S) gives the magnitude of vector S; norm(S) gives the unit vector for S
22     F_net = F + ball.m*g #Calculates the net force on the ball
23     a = F_net/ball.m #Calculates the acceleration of the ball
24     ball.v = ball.v + a * dt #Updates the ball velocity
25     ball.pos = ball.pos + ball.v * dt #Updates the ball position
26     spring.axis=ball.pos - holder.pos #Updates the spring
27     t += dt #Adds dt to the time
28
29     f1.plot(pos=(t, ball.pos.y)) #Graphs the displacement vs time
30     #f2.plot(pos=(t, F_net.y))
```

Figure 6.4 VPython code for simple harmonic motion simulation

The code is divided into three main sections. The first sets up the graph and the second sets the initial values of relevant variables. The third section updates the position velocity, acceleration and force acting on the ball. Note that the position and hence v, a and F are all treated as vectors. The output of the code is illustrated in Figure 6.5.

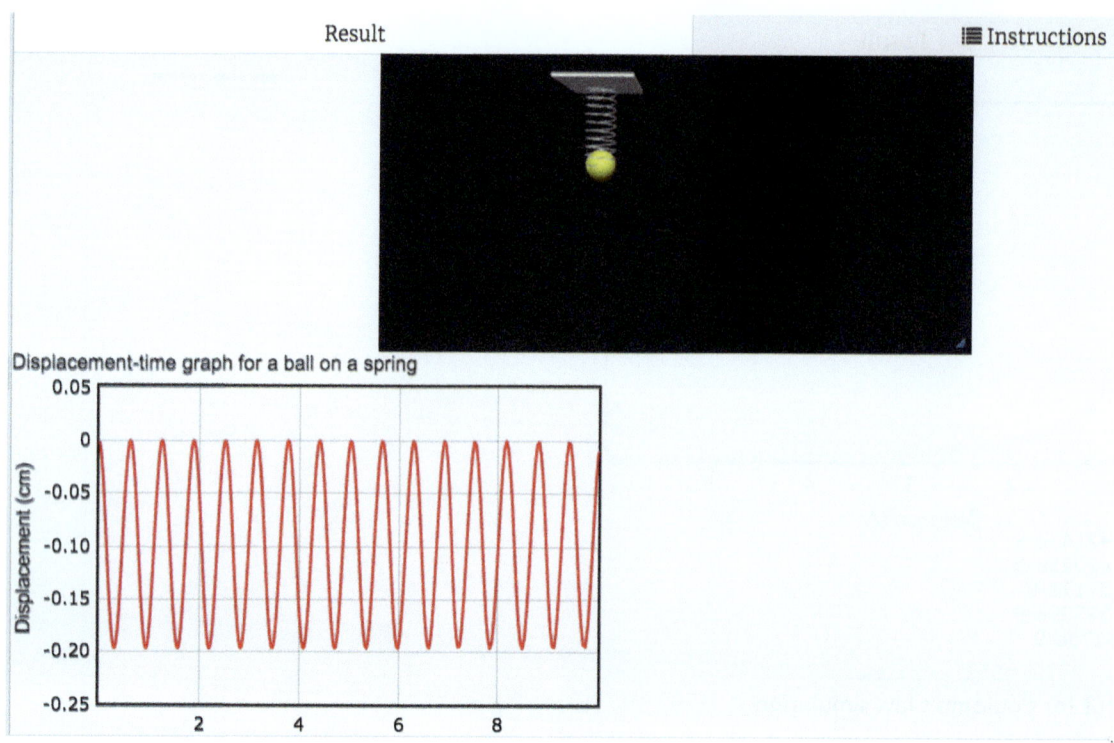

Figure 6.5 Output for simple harmonic motion simulation

Data logging

When carrying out experimental work in physics it is often necessary to either collect large amounts of raw data over a very long time period (very time consuming) or collect raw data over extremely short time intervals (very difficult).

Both of these situations can be solved using a data logger and a probe (sensor). The raw data collected by the data logger (Figure 6.6) can usually be stored in a spreadsheet, which can then be used to generate charts and graphs, although most data-logging software will allow for graphs to be generated directly from the raw data collected.

While the type of data logger available will vary, they all operate on the same basic principle. A sensor measuring a variable (for example, temperature) produces an analogue signal which the computer can convert to digital signals for processing. The data logger acts as an interface between the sensor and the computer.

Common sensors available for use in physics include those that measure electric current, pressure, potential difference, sound intensity, distance, radioactivity, light level, temperature, infrared radiation, position, magnetic flux density, ultraviolet radiation and motion.

> **Expert tip**
>
> The number of readings recorded per second is termed the sample rate.

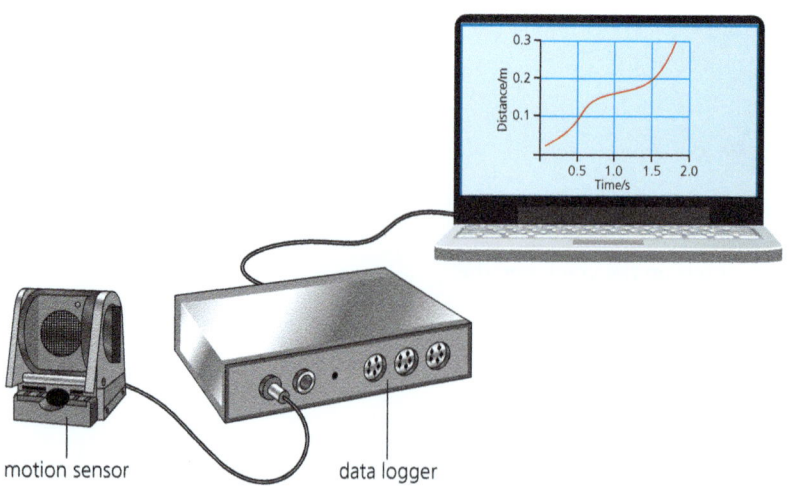

Figure 6.6 A motion sensor connected to a data logger connected to a computer

Data loggers produced by Vernier are widely used. Vernier lab manuals have basic modules that can be extended for the purposes of IAs.

Most data loggers are more than just interfaces and can store data to be retrieved at a later time. This allows them to be used remotely from a computer, which makes them ideal for field studies. Data loggers also incorporate displays that give numerical and graphical information on the readings being made in real time.

RESOURCES

https://www.vernier.com/news/category/subject-area/physics/

Expert tip

Once the raw data have been collected and a graph generated by the software, a process of inquiry should begin. For example, some or all of the following questions may be appropriate:

- For each part of the graph, what was happening during the investigation?
- What caused that peak?
- What are the highest and lowest values?
- How large was a particular change and how long did it take?
- How quickly are the values changing?
- What is the underlying **trend**?
- How does one variable seem to depend on another?
- Are there any obvious outliers?

Key definition

Trend – The general relationship shown by a set of related measurements.

■ ACTIVITY

4 Consider an experiment to investigate how the current changes with time as an electrolytic capacitor is discharged. A digital data logger and current sensor are used instead of a digital ammeter and stopwatch. State three advantages of using a data logger.

5 Calculate the number of measurements recorded in 0.5 minutes by a data logger that is sampling at a rate of 0.50 Hz.

Expert tip

Occasionally digital instruments such as a data logger will introduce an uncertainty known as a quantization error. This is caused by the conversion of an analogue (continuous) signal to a digital representation of that signal. A number of measurements are recorded (sampled) over a time interval. The signal source may vary during sampling (Figure 6.7) and the true signal and the measured signal may differ, particularly if the source signal varies at a higher frequency than the sampling frequency.

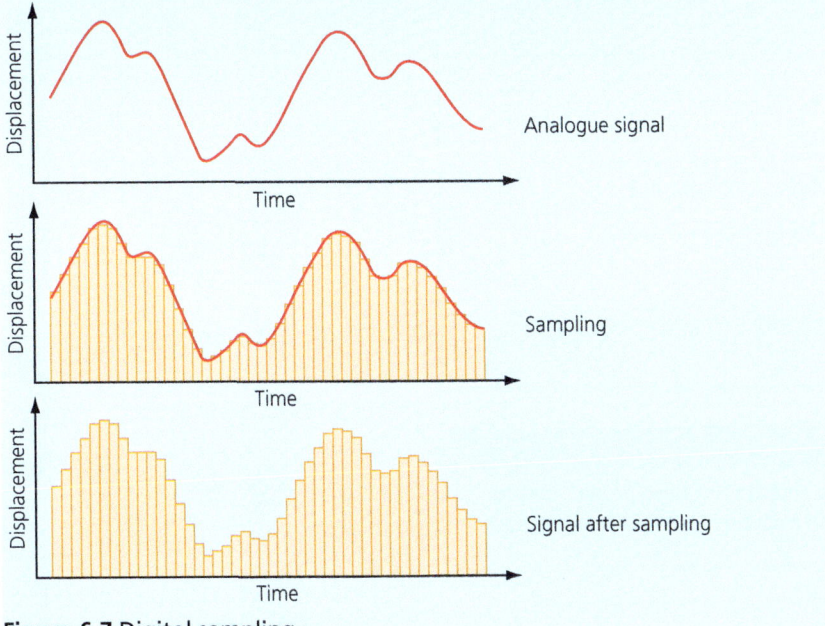

Figure 6.7 Digital sampling

6 Information communication technology (ICT)

Use of smartphones

The smartphone is a powerful computer, containing an operating system (typically Android or iOS) which allows software programs (applications, or 'apps') to be run on it.

The internet connectivity and the audio and video recording functions are just a few of the standard features of smartphones that can be used in physics investigations.

There are many science-specific apps available for smartphones; many are free and many others are available at reasonable prices. Apps can be used to measure and analyse different physical phenomena detected by the smartphone's range of sensors. Most smartphones have a microphone, as well as an acceleration sensor, a magnetic field strength sensor, a light sensor and a GPS receiver. As a result, smartphone sensors and apps can be used in investigations in fields as wide-ranging as acoustics, mechanics, optics, electricity and magnetism.

One advantage of this approach is that it allows you to record data outside the lab.

The video function enables the recording and documentation of experiments with little effort; you can then watch the videos to analyse the experiment. The recorded video sequences (for example, of a free-falling steel ball) can be analysed using appropriate software. You can measure time by counting frames – looking at the image frame by frame; for example, the Samsung S8 records at 60 frames per second. If the motion took 30 frames, then the time duration was 0.5 s.

> **Expert tip**
> Some of the free apps lack the necessary tools to assess the accuracy of their measurements.

How to obtain apps

Apps can be downloaded from online stores. Online search engines, such as Google, enable you to locate suitable apps, for example, by searching for 'physics apps + college'. Searching for 'Google Play' brings up Android apps. The Apple Store is a platform for the distribution of iOS mobile apps.

> **RESOURCES**
>
> Smart phone applications:
> - https://itunes.apple.com/us/app/galactica-luxmeter/id666846635?mt=8
> - http://journals.sfu.ca/onlinejour/index.php/i-jim/article/viewFile/3873/3206
> - https://itunes.apple.com/us/app/lab4physics/id1049405068?mt=8
> - https://phyphox.org
> - https://itunes.apple.com/us/app/vernier-graphical-analysis-gw/id522996341?mt=8
> - https://www.educationalappstore.com/app/category/physics-apps
>
> Smart phones and physics investigations:
> - https://arxiv.org/pdf/1804.06243.pdf
> - https://www.aapt.org/Resources/upload/PTE000182.pdf
> - http://smarterphysics.blogspot.com/p/publicaciones.html
> - http://www.seipub.org/fs/paperInfo.aspx?ID=9682
> - http://www.rrp.infim.ro/2014_66_4/A30.pdf
> - https://sciencejournal.withgoogle.com/

> **Common mistake**
> Be careful, because some apps display graphs that do not have proper axes for the independent and dependent variables and exclude important features such as units.

Databases

Online databases are usually good sources for interesting, complex data. For a database investigation you would access online databases to retrieve scientific information and physical data. You would design a methodology to answer your research questions using the databases, and perhaps analyse, graph or model the results.

You should investigate several reliable online sources for information, and choose the most appropriate for your IA, but you must understand and explain how these data were collected and the errors propagated.

> **Expert tip**
>
> Types of data commonly found in online databases include: astronomical data on orbits, brightnesses, velocities; particle accelerator data; nuclear data, for example average binding energies per nucleon; specific heat capacities, energy densities and heat of combustion of fuels.

RESOURCES

Online databases with physical data:
- http://hyperphysics.phy-astr.gsu.edu/hbase/pertab/pertab.html
- http://en.wikipedia.org/wiki/Energy_density
- http://www.solarsystem.org.uk/data.html
- http://www.astro.washington.edu/labs/clearinghouse/labs/labs.html

An IA may be a hybrid of 'hands-on' experimental work combined with spreadsheet/database/programming or simulation work to support and complement the laboratory work.

Simulations

A computer simulation can be used to carry out or to complement your IA investigation, provided it is interactive and open-ended. In order to use a simulation for an IA, you must design a method that allows the control of each identified independent and controlled variable that affects the dependent variable. Ideally you should obtain information or data that will be processed to make your own discovery that goes beyond the simulation's routine.

You must investigate several simulations and justify why you chose a particular simulation.

The software should allow you to estimate and propagate uncertainties.

RESOURCES

Physics simulations:
- http://www.physicslab.co.uk/gravity.htm
- https://phet.colorado.edu/en/simulations/category/physics
- https://phet.colorado.edu/en/simulation/photoelectric
- http://vsg.quasihome.com/interf.htm
- https://www.walter-fendt.de/html5/phen/singleslit_en.htm
- http://lite.bu.edu/spex/v3/index.html

Writing the Interna[tional]

Personal engagement
- Selecting an appropriate investigation
- Personal input and initiative

Writing the IA report

Analysis
- Recording a presentation of raw data
- Data processing
- Presenting data – graphing
- Impact of measurement uncertainty
- Interpreting processed data

Assessment report

Exploration
- Planning
- Variables
- Background information
- Methodology
- Safety, ethical & environmental issues
- Risk assessments

Evaluation
- Conclusion
- Strengths and weaknesses of the investigation
- Limitations of the data and sources of error
- Improvements and extensions

Communication
- Structure and clarity
- Relevance and conciseness
- Terminology and conventions
- Referencing
- Report format
- Academic honesty

7 Personal engagement

This criterion assesses the extent to which you engage with the individual investigation and make it your own. Personal engagement may be recognised in different attributes and skills. These could include addressing personal interests or showing evidence of independent thinking, creativity or initiative in the designing, implementation or presentation of the investigation.

Mark	Descriptor
0	The student's report does not reach a standard described by the descriptors below.
1	The evidence of personal engagement with the exploration is limited with little independent thinking, initiative or creativity.
	The justification given for choosing the research question and/or the topic under investigation does not demonstrate personal significance, interest or curiosity.
	There is little evidence of personal input and initiative in the designing, implementation or presentation of the investigation.
2	The evidence of personal engagement with the exploration is clear with significant independent thinking, initiative or creativity.
	The justification given for choosing the research question and/or the topic under investigation demonstrates personal significance, interest or curiosity.
	There is evidence of personal input and initiative in the designing, implementation or presentation of the investigation.

Table 7.1 Mark descriptors for the personal engagement criterion © IBO 2014

Ideally for your individual investigation you should design your own procedure. It could be inspired by an observation, an issue, or subject area of personal interest. Personal engagement can be demonstrated in:

- your research for background information
- your perseverance while collecting raw data under difficult circumstances (for example low sensitivity of readings or variables that are hard to control)
- your choice of methods of analysis (exploration).

When considering this criterion you should also bear in mind the following points:

- Your report should have a statement of purpose.
- You need to demonstrate the connection of your research question to the real world.
- The design of the methodology should be justified and show originality.

You should not use a classic investigation (that is, a well known and published method) and make little or no attempt to modify it. The subject for the investigation must be original in some way. Personal input can be reflected at the simplest level by completing the investigation, but will not score highly if you simply follow a classic experiment. You need to show you understand and can apply methodology and that you can troubleshoot and modify it, as necessary. There must be some indication in your report that you showed personal commitment to the investigation.

The criterion of personal engagement is marked using a holistic approach, using the contents of the full report of the individual investigation. It will therefore overlap with components of other internal assessment criteria, for example:

- exploration, in the selection and application of analytical and graphical techniques to process and present the numerical data
- analysis, by the comments concerning the quality of the raw data or in the discussion of the results

> **Examiner guidance**
>
> Performing an investigation with a standard method and standard analysis but in a competent way is likely to earn 1 mark for personal engagement.

- evaluation, through the depth of understanding shown in the limitations of the investigation and the reflective comments on improvements and extensions to the investigation.

The following guiding questions may help you to develop a plan for your individual investigation and to incorporate personal engagement:

- What physical phenomenon, process or system are you going to investigate?
- Which underlying theories are relevant?
- Why is it worthwhile or justified to investigate this?
- Why are you personally interested?
- Are there opportunities to show personal engagement skills, such as independent research and thinking?
- What mathematical, graphical and analytical skills do you need to apply?
- What is already known in the physics literature?
- What new physics knowledge is currently being investigated in this area?
- Is your research question answerable within the constraints of time and resources in your physics lab?
- What method will you use and adapt – will you apply an established methodology to a new topic, or apply a new methodology to an established topic?
- Can the investigation be organized into a sequence of experiments?
- How many experiments will you conduct? How long will each experiment take?
- Do you have enough time to do all the experiments in the lessons and complete the IA in an overall time of 10 hours?
- How will you record and organize your raw data and observations?
- Will you use a data logger and probes to record some raw data?
- What other apparatus, instruments and techniques will you need? Are they safe to use (with suitable safety precautions)?
- Do you know how to operate any specialist apparatus required, such as a travelling microscope, CRO, interferometer, central force frame, Millikan's apparatus, Coulomb's law apparatus, Atwood machine, or spectrometer?
- How will you control and monitor the controlled variables?
- Can you find relevant secondary data?
- Can you simulate some aspects using Excel or Python?

Many ideas for IAs are formed when a student extends their understanding of one area of the curriculum that they were interested in. For example, you may be familiar with the discharge of a capacitor through a resistor to measure the time constant. The transient experiment can be done by using a voltmeter and stopwatch, signal generator and oscilloscope. However, you may plan an investigation studying the alternating current across the capacitor and the resistor as a function of frequency. This leads into the area of impedance.

Or, you may have investigated the current and pd characteristics displayed on an oscilloscope of a sine wave from a signal generator. A suitable investigation to develop from this could be to electrolyse dilute sulfuric acid with aluminium or platinum electrodes at high ac frequencies.

You may wish to explore topics outside the current IB Diploma Programme physics course, such as the MOFSET and Zener diode (electronics).

Examiner guidance

Do not forget that the total time allocation for your IA is 10 hours – this includes planning time, implementation, and writing of the report.

RESOURCES

Ideas for practical-based investigations can be found at:
- http://fep.if.usp.br/~profis/projetos/NUFFIELD/Revised_Nuffield_Advanced_Physics/Examinations_Investigations.pdf
- https://www.talkphysics.org/articles/physics-education-papers-a-level-practicals/
- https://obelkobusnel.files.wordpress.com/2012/03/300-lab-ideas.pdf
- http://aapt.org
- http://www.iop.org
- http://newt.phys.unsw.edu.au/~jw/l&lexperiments.pdf
- https://www.qcaa.qld.edu.au/downloads/senior/snr_physics_07_eei_ideas.pdf

Examiner guidance

You must cite all references that relate to the development of your methodology.

RESOURCES

Ideas for investigations involving electronic components:
- http://www.electronics-tutorials.ws/transistor/tran_7.html
- http://www.antonine-education.com/jirvine/Electronics/EL_Intro/Basic_measurement/intro_page_2.htm
- http://www.antonine-education.com/jirvine/Electronics/EL_Tut_04/tutorial_4_diodes.htm
- http://www.electronics-tutorials.ws/diode/diode_7.html

Justifying your research question

The topic of study may be of significance to a particular community, for example, the harmonics in a guitar string to musicians, or to the economy, for example, the efficiency of photovoltaic cells, or to the environment, for example, the effect of shelving on the breaking of waves.

Your research question could be based on an observation of an interesting demonstration. For example, you may have seen a magnetic cannon or Gauss rifle, which converts magnetic energy into kinetic energy. (When a steel ball with low initial velocity impacts a chain consisting of a magnet followed by additional steel balls, the last ball in the chain gets ejected at a much larger velocity. The analysis of this device involves an understanding of advanced magnetism, energy conversion, and the collision of solids.)

Or, your observations in everyday life may have given rise to a question. For example, when placing a ladder against the wall, what is the maximum angle from the wall to the ladder to ensure the ladder will not slip? This could lead to an investigation to determine the relationship between angle and coefficient of friction, supported by a mathematical analysis. (See 'The ladder against the wall revisited', *School Science Review*, March 2016, 97(360), 84–89.)

You must not repeat and extend a practical you have done previously, unless you use an unrelated technique (instrument and/or analysis) or you apply the same technique in an unfamiliar area of physics.

> **Common mistake**
>
> Do not assume that 'personal engagement' means 'personal significance'. Avoid contriving personal significance of your investigation, for example, by simply stating 'I have always been interested in…', without giving further concrete reasons for your choice. Neither should you write long artificial comments about your interests and personal situation.

Evidence of personal input

There needs to be evidence of independent thinking, personal input and initiative in the design of your investigation and/or in its implementation. This can be shown through the level of commitment you show throughout the whole process, including persistence in collecting raw data, the design of apparatus, data analysis and presentation or modification of techniques.

> **Expert tip**
>
> You do not need to exhibit particular skills beyond the scope of the IB Physics course. For example, should you decide to investigate some aspect of the physics of a pinhole camera (for example, investigating the angular resolution of the pinhole camera as a function of hole size), you do not need to be a photographer. There are claims that pinhole glasses are better than conventional lenses for correcting refractive defects, such as myopia, which connects the investigation to the real world. (See 'Optics in the pinhole camera', *School Science Review*, June 1994, 75 (273), 59–66.)

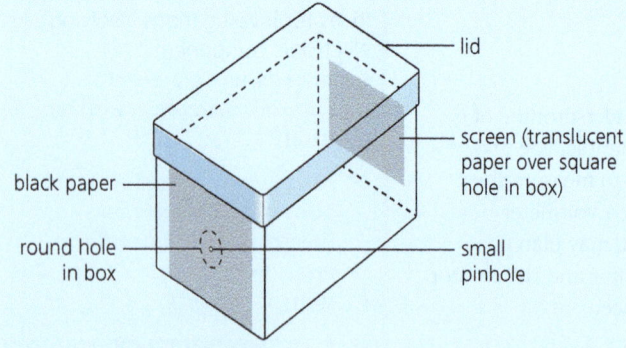

Figure 7.1 Pinhole camera

Your stated research question must connect with your method. For example, if your research question suggests investigating the effect of speed of electrons but your method actually investigates the speed of diffusion of coloured ions in solution, then this would be penalized. (The charge carriers in ionic solutions are ions and not electrons.)

> **Examiner guidance**
>
> You may be likely to score highly in this criterion if you design and construct your own apparatus. Examples could include a Kundt's tube (for measuring the speed of sound in air and gases, significant to aerodynamics and aircraft design), a polarimeter for analysing optical activity (used in determining the quality of sugar products), a wire or current balance for illustrating the Lorenz force and measuring magnetic fields, an interferometer, or a hot-wire ammeter (Figure 7.2), a historic instrument measuring ac and dc based on the heating effect of a current. However, such an approach, if very time consuming, may be more appropriate for your extended essay.
>
>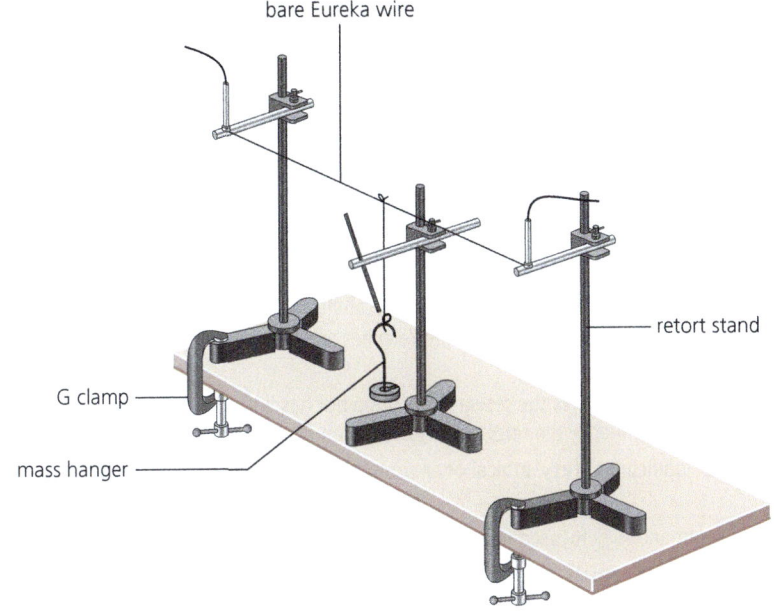
>
> **Figure 7.2** One way of modelling a hot-wire ammeter

Personal engagement criterion checklist

■ Creativity input and initiative

Descriptor	How?
You demonstrate creativity during the individual investigation (design, implementation and presentation).	
You demonstrate independent thinking during the entire individual investigation (design, implementation and presentation).	
You demonstrate initiative during the entire individual investigation (design, implementation and presentation).	

■ Justification for research question

Descriptor	Complete
You justify why you chose to investigate your research question and include your personal significance, interest and curiosity.	
You discuss the wider importance and significance of the research question.	

8 Exploration

This criterion assesses the extent to which you establish the scientific context for the work, state a clear and focused research question and use concepts and techniques appropriate to the level of the IB Diploma Programme physics course. Where appropriate, this criterion also assesses your awareness of safety, and environmental and ethical considerations.

Mark	Descriptor
0	The student's report does not reach a standard described by the descriptors below.
1–2	The topic of the investigation is identified and a research question of some relevance is stated but it is not focused.
	The background information provided for the investigation is superficial or of limited relevance and does not aid the understanding of the context of the investigation.
	The methodology of the investigation is only appropriate to address the research question to a very limited extent since it takes into consideration few of the significant factors that may influence the relevance, reliability and sufficiency of the collected data.
	The report shows evidence of limited awareness of the significant **safety**, ethical or environmental issues that are relevant to the methodology of the investigation.*
3–4	The topic of the investigation is identified and a relevant but not fully focused research question is described.
	The background information provided for the investigation is mainly appropriate and relevant and aids the understanding of the context of the investigation.
	The methodology of the investigation is mainly appropriate to address the research question but has limitations since it takes into consideration only some of the significant factors that may influence the relevance, reliability and sufficiency of the collected data.
	The report shows evidence of some awareness of the significant **safety**, ethical or environmental issues that are relevant to the methodology of the investigation.*
5–6	The topic of the investigation is identified and a relevant and fully focused research question is clearly described.
	The background information provided for the investigation is entirely appropriate and relevant and enhances the understanding of the context of the investigation.
	The methodology of the investigation is highly appropriate to address the research question because it takes into consideration all, or nearly all, of the significant factors that may influence the relevance, reliability and sufficiency of the collected data.
	The report shows evidence of full awareness of the significant **safety**, ethical or environmental issues that are relevant to the methodology of the investigation.*

*This indicator should only be applied when appropriate to the investigation.

Table 8.1 Mark descriptors for the exploration criterion © IBO 2014

Selecting a research question

A suitable individual investigation may involve studying how an independent variable (numerical and continuous) causally affects a single dependent variable, while a range of other variables are controlled or at least monitored. Typically a relationship is established between an independent variable and a dependent variable. The practical work may be supported by secondary data and a simulation.

The independent variable and its range need to be stated and justified, and all other variables identified with their methods of measurement or their control. Trials may be used to determine appropriate values for an independent variable. The discussion of controlled variables demonstrates that you understand that other factors may impact on values of the dependent variable (that the investigation should be a **fair test**).

In addition to a focused research question, the exploration criterion will assess the presentation of background information that provides context and reasons behind the investigation. This needs to be focused and contain only relevant information.

Your IA report needs to describe and explain the safety, ethics and environmental impact of the investigation, where appropriate.

> **Key definition**
>
> **Fair test** – A focused experiment, adhering to the scientific method, in which only the independent variable is allowed to significantly affect the dependent variable.

> **Examiner guidance**
>
> A drawback of a simple experiment is that its findings may not be applicable to a wider setting.

Critical questions to ask when formulating your research question and methodology may include:

- Is there a single and well defined independent variable and a quantifiable dependent variable?
- Does the methodology include physical measuring techniques, require a detailed mathematical analysis and allow control of variables?
- Can I construct the apparatus and obtain a set of rough measurements within a few hours?
- Am I likely to generate meaningful raw quantitative data that can be processed and used either to support or falsify my research question?
- Are the underlying physical models and numerical relationships accessible to me?

If some of your answers to the above questions are 'no' then you should consider rejecting the investigation topic.

> **Expert tip**
>
> An investigation that begins with the aim of investigating a simple physical phenomenon and then becomes more complicated is not a 'failure', even if several questions are unresolved.

> **Investigations**
>
> It is difficult to start a balloon inflating but after a pressure peak it becomes easier. A suitable practice investigation may be to establish the experimental pressure versus radius graph (using the barometer of a smartphone) for a rubber balloon and to compare it to the theoretical one. (See 'On the inflation of a rubber balloon', J. Vandermarlière, *The Physics Teacher*, Volume 54, December 2016.)

Investigations may be possible using *items at home*, for example, resonance effects in bottles or glasses, or the characteristics of loudspeakers. You may also consider using sports or vehicles as a stimulus for developing an investigation, for example, the load/speed variation of parachutes, the effect of internal pressure on the performance of a football, the action of sails or the flight properties of a golf ball.

Apparatus present in a laboratory may also be the starting points for an investigation, for example, the Pitot tube, Venturi meter, cloud chamber, Fresnel lens, polarimeter, dry cell, Schlieren photography apparatus or van de Graaff generator. Certain physical effects that interest you may provide the 'spark', for example, eddy currents, Newton's fringes or the piezoelectric effect.

Stimuli for investigations may also come from observing natural phenomena, for example, corona discharge, creep in copper wire, crater formation, the behaviour of gas bubbles rising through liquids or vibrations in a hacksaw blade (Figure 8.1). For the latter, a possible research question might be: how does the period of oscillation of a cantilever (fixed at one end) depend on its length?

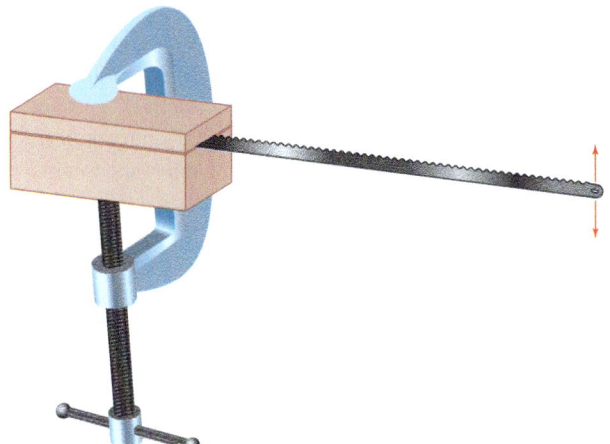

Figure 8.1 A vibrating hacksaw blade

The IB learner profile encourages IB students to be 'risk takers' and this can be interpreted as encouraging you to investigate physical relationships not included in the IB Diploma Programme physics course. You could, for example, determine the temperature (by heating with direct current) at which the magnetic properties of an iron, nickel or cobalt wire change. This is known as the Curie point. You could even investigate whether tension or the use of an alternating current affects the result.

You may be enthused to investigate the gyroscopic effect of angular momentum after watching Mike Fossum demonstrating this in the International Space Station. (See **https://www.youtube.com/watch?v=2Oc-Ucx_4Ug**.)

> ### Investigations
>
> You could consider measuring Young's modulus (Figure 8.2) for various materials, including wood. This is not in the current IB Physics syllabus so it needs to be defined. The Young's modulus of a material is the ratio of stress to strain (that is, the measure of resistance to elastic deformation).
>
>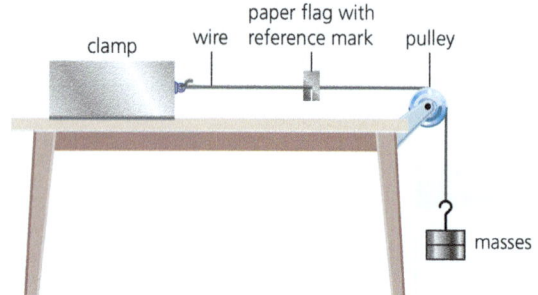
>
> **Figure 8.2** Simple experiment to measure the Young's modulus of wire
>
> You should link the concept to engineering, since it is a material property that describes the material's tensile elasticity along an axis when opposing forces are applied along that axis. It is one of the most important properties in engineering design.
>
> Young's modulus is not always the same in all orientations of a material. Most metals and ceramics are isotropic, meaning their mechanical properties are the same in all orientations. However, anisotropy can be seen in some treated metals, many composite materials, wood and reinforced concrete.

Even a simple observation such as a ball rolling down an inclined plane onto a table top (Figure 8.3) can be analysed theoretically and compared to experimental results. (See 'A simple and surprising experiment is performed by physical science students', M. Lattery, in *Physics Education* 35(2): 130–131, March 2000.) A relevant research question might be: for what angle will the time of travel across the table top be a minimum?

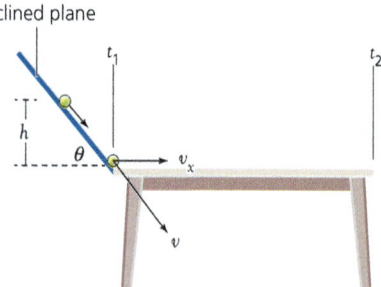

Figure 8.3 Investigating travel time of a rolling ball across a table top

Comparisons of techniques or of instruments could form good investigations. Sometimes there is more than one way of recording measurements for determining a physical constant. For example, the acceleration due to gravity, g, can be measured by timing accurately the fall of an object over a known distance, by using a pendulum, or by measuring the gain in velocity of a falling object during a measurable time. Deciding which method generates more accurate and reliable data will involve quantifying the random uncertainties and systematic errors involved in each case.

A thermistor and a thermocouple could be compared as calibrated thermometers.

The measurement of the depth of liquid in a container using a float attached to a rotary potentiometer could be compared with the use of an LDR at the bottom detecting attenuation from a light source at the top.

> **Examiner guidance**
>
> You should not use a mandatory practical or a simple practical, such as investigating resistors in parallel or in series, as the basis for your IA. But you could adapt and extend one; for example, determining the value of the spring constant by a static method would be too simplistic but an investigation (using a dynamic method) of how temperature or alloy composition affects the spring constant might be a suitable one.

> **Investigations**
>
> Consider the suggested IB practical: determination of the thickness of a pencil mark on paper (which can be estimated from its electrical conductivity). The value of the resistivity of pencil 'lead' (assumed to be the same as that of graphite) can be used to estimate the thickness of a pencil line from its width, length and resistance.
>
> You should first do some rough calculations to obtain an estimate of the resistance of a pencil line. Use this value as a guide to selecting the power supply and meters.
>
> The challenge of making electrical contact with the pencil line and obtaining reproducible and reliable results needs careful consideration. Allowance could be made for contact resistance by comparing the current passed by two similar pencil lines but of different lengths.
>
> Accuracy could be improved by measuring the resistance and hence the resistivity of the pencil lead itself.

Phrasing the research question

The 'research question' may be phrased as a focused scientific *question* but this must be one that is answered with a statement, not just by a 'yes' or a 'no'. For example, 'How does varying the length of a simple pendulum of fixed mass affect its period of oscillation?' is much better than 'Does the length of a pendulum affect its period?'

> **Expert tip**
>
> There are several types of pendulum that you could investigate: simple, torsional, compound and bifilar (Figure 8.4). Specify the type in your research question.

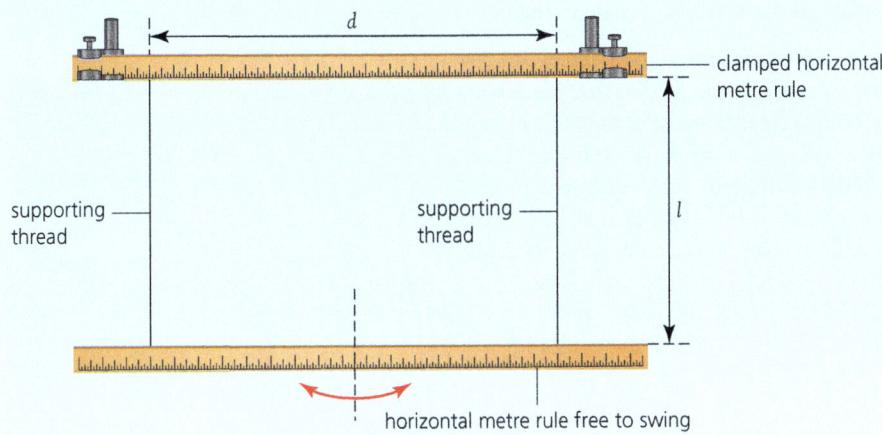

Figure 8.4 Bifilar suspension

The research question must specifically link the independent and dependent variables. For example:

- How does the temperature of a squash ball affect its vertical bounce height from a smooth, hard surface?
- How does the linear acceleration of a steel ball down a wooden ramp depend on the height of the ramp?
- How does varying the angle between a search coil and magnetic field direction affect the measured magnetic flux density?
- How does the rate of flow of a fixed mass of water (at constant temperature) from a siphon depend on the diameter of the outlet tube?
- What is the relationship between the depth of a crater (in fine sand) and the drop height of a steel ball bearing?

> **Examiner guidance**
>
> The question 'What factors affect the blowing of a fuse?' is not focused, since several factors affect the size of the current at which the fuse will blow, including the material of the wire and the diameter of the wire.

Examiner guidance

The research question should *not* be stated in the form of verifying a law or well known relationship. For example, an investigation to verify Newton's law of cooling (Figure 8.5) could be phrased in the form of a suitable research question as 'How does the rate of cooling of hot water (stirred by a magnetic stirrer) in an open glass beaker vary with the surrounding temperature?'

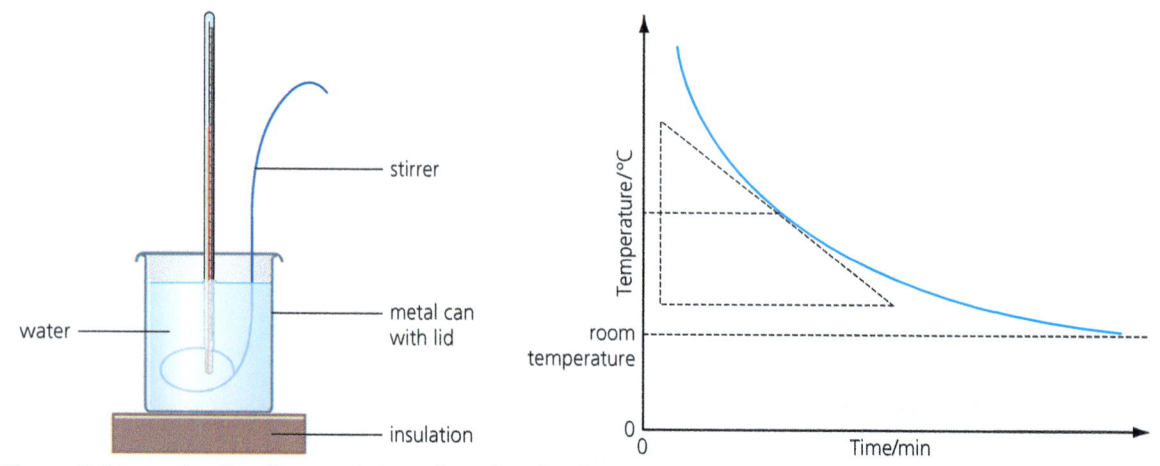

Figure 8.5 Investigating Newton's law of cooling (under conditions of forced convection)

A 'research question' can, alternatively, be stated as the *objective* or *aim* of the investigation. For example, 'To determine and compare the theoretical and experimental values of the moment of inertia for a ping-pong ball.' This must also identify the independent variable and the dependent variable. It may also mention the method. For example: 'To discover the relationship between the intensity of plane-polarized light and the angle between the polarization axis of the analyser and the axis of polarization, as measured by the light sensor of a smartphone.' Here, the independent variable is the angle and the dependent variable is the intensity of emerging light.

Consider the research question 'How does the linear acceleration of a steel ball down a ramp depend on the height of the ramp?' You are likely to predict that when you increase the height of the ramp, then you will increase the acceleration of a ball which rolls down the ramp. This prediction will need an explanation framed in terms of components of forces (illustrated by a free body diagram, such as Figure 8.6) and work and energy (kinetic and gravitational potential). This simple model for a trolley rolling down a slope assumes sliding, not rolling, which would complicate the issue.

ACTIVITY
1. Look up the *Mpemba effect* and the *Kaye effect* on the internet. Suggest a suitable research question for each effect.

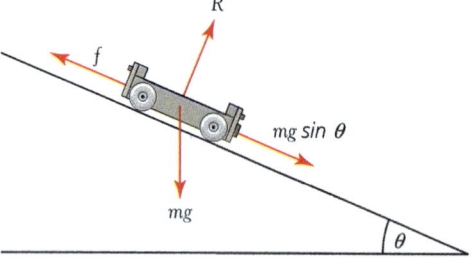

Figure 8.6 Free body diagram for a trolley rolling down a slope

Background information

You need to include in your IA report a review of the physics literature (which must be fully referenced) related to your individual investigation. This has the following functions:

- to justify your choice of research question and methodology
- to establish the importance and significance of the topic
- to provide relevant background information needed to understand the report.

The background will provide a brief overview of the theory and current knowledge, with an emphasis on the physics literature specific to the topic. It should support the argument behind your report, using evidence from that research area.

Expert tip
Do not include any advanced theory that is not directly relevant, or you do not understand and cannot explain clearly.

■ Constructing a hypothesis

A research question may be used to formulate a hypothesis and this may be a useful inclusion in your background information, but it is not an explicit requirement of the exploration criterion in the group 4 IA.

A hypothesis is a testable prediction and explanation of the type of physical behaviour or result expected. Hypotheses enable the design of investigations so that further predictions based upon the hypothesis may be tested and either tentatively supported or falsified.

Where possible, the hypothesis should represent a quantitative, mathematical relationship between two variables. For example: 'When mass is kept constant, the acceleration on an object (on an air track) produced is directly proportional to the applied force.' Any assumptions should be stated (for example, absence of friction is assumed) or explained (for example, an ideal liquid implies zero compressibility and zero viscosity).

> **Worked example**
>
> The speed of sound, v, in a gas of density, ρ, is given by the relationship:
>
> $$v = \sqrt{\frac{\gamma p}{\rho}}$$
>
> where p = pressure and γ is a constant. Therefore, in an experiment to verify this, you might hypothesize that if the pressure and density of a gas are multiplied by the same factor, no change will occur in speed. If the pressure of a fixed mass of gas at constant temperature is doubled, then its volume will be halved according to Boyle's Law so the density will be doubled – and therefore the speed will be unchanged. The speed of sound is independent of pressure changes at constant temperature (assuming ideal behaviour).

Variables

Variables are factors that can be measured and/or controlled during your investigation. An independent variable is changed, and the result of this change leads to a change in the measurement of the dependent variable. A controlled variable is one that should be held constant, to simplify determining the effects of the independent variable on the dependent variable.

The variables in your investigation need to be explicitly identified by you as the dependent (measured), independent (manipulated) and all relevant controlled variables (constants). Relevant variables are those that can reasonably be expected to affect the values of the dependent variable.

For example, consider the research question, 'To determine the relationship between the period of oscillation of a loaded ruler and the overhanging length of the ruler (for a fixed mass)'. The setup for this is shown in Figure 8.7.

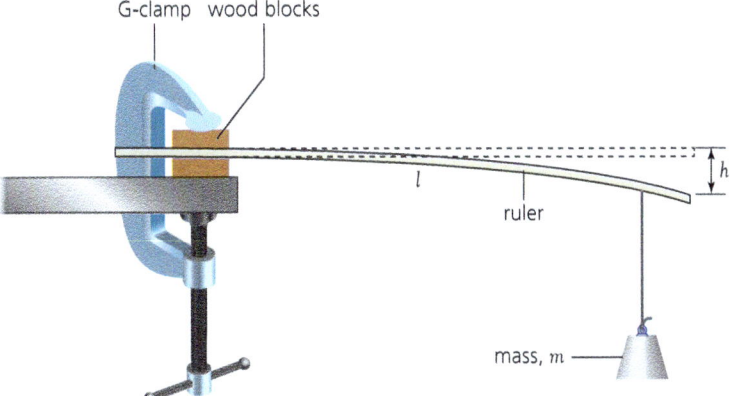

Figure 8.7 An oscillating cantilever

You must clearly state that:

- the independent variable is the overhanging curved length, l
- the dependent variable is the mean period of oscillation
- the controlled variables are the mass of load at the end of the ruler, the position of the load, and also the physical properties of the ruler (its modulus of elasticity, its width and thickness, achieved by using the same ruler during the investigation).

Methodology

In your investigation you must record relevant, reliable and sufficient raw data in order to address the research question. *Relevant* means the data are related to your research question; *reliable* means precise and accurate; *sufficient* means a wide range of repeated data is collected. Enough repetitions should be planned and performed to establish the basic reproducibility of the results, plus a wide enough range/sufficient values of the independent variable to answer the research question.

The classified variables, including the controlled variables, could be presented in a table together with the planned method of their measurement or control. Table 8.2 shows one suggested format. The example is a simple investigation to determine the effect of a change in temperature on the volume of air at constant pressure (verifying Charles' law), as shown in Figure 4.11.

Type of variable	Variable	Method of measurement or control	Reason for control
Independent	Temperature of gas	The tube with air will be placed in a water bath at a temperature which can be adjusted to a range of different constant values.	
Dependent	Volume of air	The length of the air column will be measured for a tube of known internal cross-sectional area.	
Controlled	Pressure of air	It can be monitored using a barometer at the beginning and end to establish it has not altered within the precision of the instrument.	Any pressure change at constant temperature will change the volume of the trapped air.
Controlled	Mass of air	The apparatus will be weighed before and after the experiment on a very accurate electronic balance to ensure a constant mass of gas and that the tube is sealed.	A decrease in the mass of the gas will decrease the pressure of the gas.

Table 8.2 A possible format for presenting information about the variables

You should provide enough detail of your procedure so that another IB Physics student could repeat your work without your presence. This means giving a detailed method, a simple, labelled 2D diagram drawn in cross section (or a labelled digital photograph), a list of apparatus and instruments with their tolerance (manufacturer's stated error), range and sensitivity.

If space is limited in your report, illustrations can be limited to complex setups, non-standard equipment or standard equipment being used in an unusual manner.

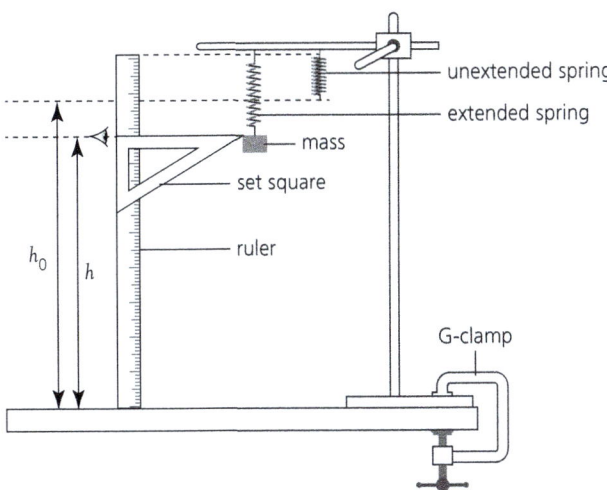

Figure 8.8 Example of a report diagram: apparatus to determine the restoring force per unit extension of a spiral spring

You should also include in your outline of the procedure any safety considerations. In the example of determining the restoring force per unit extension of a spiral spring (Figure 8.8), masses should be attached securely to the spring, the lab floor should be protected from falling masses by using a crash box or mat, and the clamp stand must be securely attached to the bench by a G-clamp.

The method can be written in continuous prose (as in a scientific paper) or as a list (in bullet points). It should not be a long list of detailed instructions, but it should explain why certain actions are performed. Any steps or procedures designed to minimize the systematic and random errors in your measurements should be clearly described. For example, for the experiment shown in Figure 8.8:

- The axis of the spiral spring must be vertical so that it extends along the plane of measurement.
- The scale should be set up vertically in a position that will allow measurement of the maximum extension.
- The spiral spring should not be stretched beyond its elastic limit.
- The mass should be placed gently and should be increased in equal steps.

The advantages and **limitations** of one type of apparatus, instrument or practical approach compared to other possibilities should be briefly outlined, where appropriate. It may be helpful to have two sections: first, the planning or design and development of the method; second, the methodology actually used for collecting data.

For example, a simple pendulum provides a simple and easy approach to determining the value of the acceleration due to gravity, g. However, it suffers from the following limitations, which may be determined from preliminary experimentation:

- In practice there is a maximum oscillation amplitude that can be achieved so the deviation from the model due to the angle of the swing is relevant.
- The motion of the bob is not purely translational. It also possesses a rotatory motion about the point of suspension.
- The suspension thread has a finite mass and hence a definite moment of inertia about the point of suspension.
- The suspension thread slackens when the limits of swing are reached and so the effective length of the pendulum does not remain constant during the swing.
- Corrections are needed for the finite size of the bob, the yielding of the support and the damping due to air resistance.
- Preliminary experiments with a compound pendulum suggest that many of these limitations are absent or reduced.

> **Expert tip**
>
> It may be helpful to ask another physics student to read your methodology and then get them to explain your investigation to you. This will help ensure your approach is understandable and reproducible.

> **Key definition**
>
> **Limitations** – The restrictions of a particular experimental technique or instrument, which may influence the results.

Safety, ethics and environmental impact

Ethics are moral principles that govern your behaviour. Ethics applied to the practice of experimental physics work may imply the following ethical rules: accurate and full investigation reports, objective analysis and interpretation of data and fair citations. Your report should contain considerations of safety, as well as ethical and environmental issues, in the form of a risk assessment. If there are no serious issues to be addressed, you should state this. The three main parts of a risk assessment are:

- **Hazard identification** – identifying safety and health **hazards** associated with laboratory work (Figure 8.9).
- **Risk evaluation** – assessing the **risks** involved.
- **Risk control** – using risk control measures to eliminate the hazards or reduce the risks.

> **Key definitions**
>
> **Hazard** – The potential to cause harm.
>
> **Risk** – The probability of harm occuring.

Figure 8.9 Hazard warning signs for chemicals, lasers, ionizing radiation and high voltage

The aim of risk assessment is to minimize the possibility of a potential source of harm becoming an actual source of harm.

Electricity and magnetism

The main hazard in a physics laboratory is mains electricity. When working with circuits make sure the power supply is off and unplugged when assembling or disassembling a circuit. Ensure the circuit is connected correctly before applying power, to prevent damage to the equipment.

If your circuit includes a rheostat, start with it set at maximum resistance. Use the largest scale on the meter then work your way down. When changing the resistance of a circuit, make sure that the resistance does not become small enough to burn out the meter.

Large permanent magnets (especially those containing rare earths) and electromagnets may attract opposite magnetic poles or steel objects with a very strong force. There is the potential hazard that fingers get pinched when caught between the magnet and the iron or steel object.

> **Examiner guidance**
>
> The following all involve high risk: high voltage power supplies, chemicals, lasers, radioactive sources, liquid nitrogen, hot plates, ultrasonic baths, machines and hand tools.

> **RESOURCES**
>
> Source of safety data in the UK:
>
> http://science.cleapss.org.uk/resource/Student-Safety-Sheets-ALL.pdf

Mechanics

Investigations in mechanics may involve rapidly moving objects (such as model rockets), falling objects or stretched springs. Being hit by fast-moving objects can produce bruises, scratches and cuts, or eye injuries. Safety glasses must be worn to reduce the latter risk.

Pre-test any projectile (which must not have a sharp point) to determine the path it will take. Ensure that no one is along the path or in the impact area, in the lab or outside. Use a simple mechanical launcher (for example, a compressed spring, compressed air, or stretched elastic). Load the launcher only when it is to be fired.

Heavy masses may be used in experiments studying free-fall, Newton's laws of motion, impulse and momentum. Care should be taken to prevent feet or hands being caught between a moving heavy mass and the floor or table surfaces. For falling objects, position a 'crash box' (a box containing a soft material that will cushion the fall of the object and prevent it from bouncing). Stationary devices should be secured by a G-clamp.

Take care to avoid unexpected release of a spring's potential energy when working with dynamics carts, spring-type simple harmonic oscillators, and springs used in wave investigations. A stretched spring, unexpectedly released, can damage fingers. A compressed spring, when suddenly released, can hit someone or send a projectile at high velocity.

Heating

If your investigation involves heating a substance, you should consider your heating method. Bunsen burners can reach high temperatures quickly but can be hard to control. They should never be used if flammable liquids (for example, alcohols and hydrocarbons) are involved.

Water baths are safe to use but are very slow to heat, cannot heat anything above 100 °C (the normal boiling point of water) and have poor temperature control. If you decide to use a heat lamp or immersion heater, take care because these can become very hot to handle and can take a long period of time to cool down. Also, they may not provide uniform heating to the substance being heated.

Report structure

Research question: To establish the relationship between the distance from a cobalt-60 source and the intensity of gamma radiation measured by a Geiger–Müller (GM) tube.

Variables: The independent variable is the distance, d, of the cobalt-60 source from the mica end window of the GM tube; the dependent variable is the count rate. The controlled variable is the intensity of the gamma radiation (the same source is used throughout the experiment; cobalt-60 has a long half-life and hence its activity can be assumed constant for the duration of the experiment). Temperature and air pressure have no effect on the rate of radioactive decay.

Safety precautions: The experimenter should maximize their distance from the cobalt-60 source and never touch it by hand. Disposable gloves should be worn and the source handled by a technician or teacher with tongs or tweezers. The source should be returned to storage in its lead-lined box. Hands should be washed after the experiment.

Methodology: The background rate should be measured without the cobalt-60 source present. The source should be placed at a fixed distance, for example, 0.10 m, from the GM tube and the distance measured. The number of counts (gamma emissions) for a fixed period, for example, 60 s, should be recorded. This should be repeated three times and the mean count calculated. This should then be repeated for at least five different distances. The mean counts should be corrected by subtracting the background count.

Hypothesis: It is expected that the data will obey an inverse square law, that is that the count rate will be inversely proportional to the square of the distance between the source and the detector.

Results and analysis: A graph is plotted of distance against $\frac{1}{\sqrt{\text{corrected count}}}$ and a line of best fit drawn.

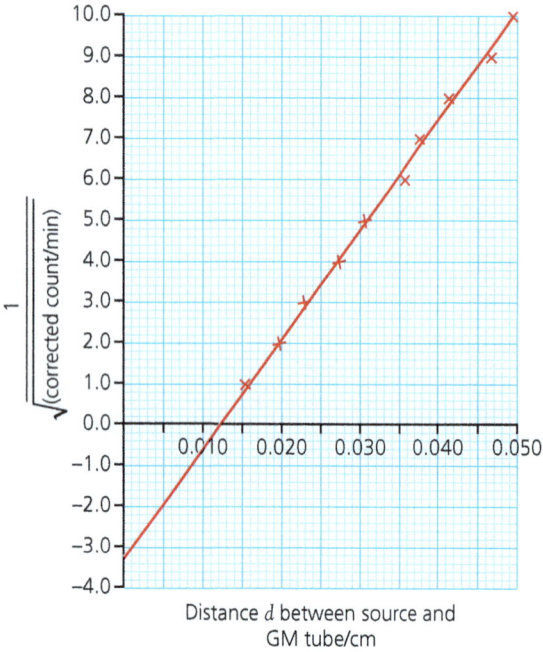

Figure 8.10 Distance against $\frac{1}{\sqrt{\text{corrected count}}}$ for a gamma-emitting cobalt-60 source

The line of best fit is straight but it does not go through the origin, so the results do not verify the inverse square law.

Background theory: The reason for the non-zero y-intercept on the graph is that the source itself is a small distance inside its container and the detection point of the gamma radiation is somewhere inside the GM tube, not at the mica window.

The true distance between the source and the place where the gamma radiation is detected is $r = d + d_0$, where d_0 is a correction factor.

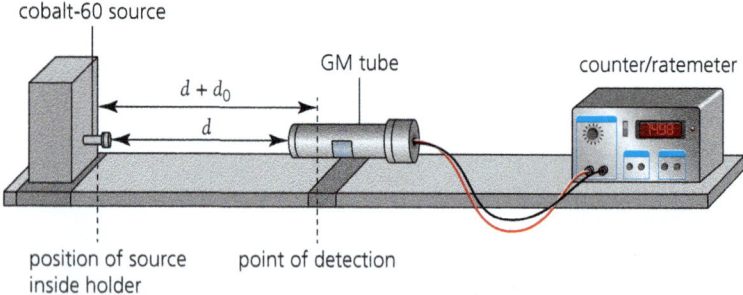

Figure 8.11 Corrected distance of point of detection from source

The intensity, I, of gamma radiation depends on the distance, r, from the source according to the inverse square law, so I is proportional to $\frac{1}{r^2}$. In this experiment, the corrected count rate, C_c, is proportional to the intensity of the gamma radiation. Therefore:

$$C_c = \frac{k}{(d+d_0)^2}$$

where k is an experimental constant.

Rearranging:

$$d = \sqrt{k} \times \frac{1}{\sqrt{C_c}} - d_0$$

This is of the form $y = mx + c$, where the gradient $m = \sqrt{k}$ and the y-intercept $c = -d_0$.

The results show that the correction factor $d_0 = -3.2$ cm.

■ ACTIVITY

2 Research on the internet suggests that the terminal velocity of a steel ball bearing falling through oil is related to its radius by the expression $v = kr^2$, where k is a constant. Design a laboratory experiment to verify this. Provide a research question, a classification of variables, an outline of a suitable method (with a labelled diagram), and how the data will be analysed. Outline any safety issues.

Exploration criterion checklist

■ Defining the problem and selecting variables

Descriptor	Complete
Research question	
You identify an appropriate topic.	
You state a relevant, specific and fully focused research question clearly, where the dependent and independent variables, and perhaps method, are included.	
Your research question sets the framework for the entire individual investigation and is consistently carried through.	
You state and describe the correct independent (or manipulated) variable with the correct range of values it will take.	
You state and describe the correct dependent (or measured variable) and processed variable(s) (with their appropriate units).	
Where appropriate, you predict a quantitative relationship between the independent variable and the dependent variable(s).	
You state and describe the relevant controlled variables together with why and how they are controlled or monitored.	
Background information	
You give detailed, relevant and appropriate physics and mathematical background theory (including mathematical derivations of formulae and explanation of symbols, where appropriate).	
Your information enhances understanding of the investigation and puts it into a physical science context.	
You include a hypothesis or physical model (quantitative or qualitative), when possible and appropriate.	
You outline any assumptions, limitations or simplifications in any physical models that you discuss.	
You include a brief survey or summary of the physical science literature that is referenced according to a stated referencing style.	
Safety, ethical and environmental issues	
Your plan shows awareness of safety, ethical or environmental issues related to the methodology, for example: risk assessment, use of materials, instruments or apparatus and any issues related to storage and disposal of chemicals or materials.	

Controlling variables

Descriptor	Complete
You include details about apparatus, materials and instrumentation (including absolute uncertainty in a single measurement and sensitivity and range).	
You include a complete description of any materials or substances and the dimensions and composition of any solids. (A single reference in the method is acceptable.)	
You give a clear, detailed and logical sequence of reproducible steps.	
You describe the rationale or justification of relevant steps in the setup of the apparatus.	
You describe how your methodology minimizes random uncertainties.	
You describe any calibration of instruments and checking for systematic errors.	
You present a cross-sectional labelled diagram (photos, where appropriate) showing the arrangement of non-standard apparatus. Correct names and terminology are used.	
You include circuit diagrams, free body diagrams and ray diagrams, where appropriate.	
You outline planned controls (if appropriate) and simple statistics, if large amounts of repeated readings are recorded.	
You explain choices with regard to the methodology, apparatus or instrumentation selected and materials or substances.	
You have explored alternative methods and outlined why they are less suitable.	

Planning and recording of raw data

Descriptor	Complete
You plan to collect a sufficient number of reliable and relevant raw data points over a wide data range.	
You plan to collect more raw data at certain points, for example, extremes of the range and inflexion points.	
You plan to collect a suitable number of repeated and averaged readings.	
You plan to collect relevant qualitative data (observations).	
Your method takes into account and minimizes likely random and systematic errors in the raw data.	
You ensure that your data collection is relevant to the initial research question.	
You plan to collect raw data that records units and random uncertainties, as well as being recorded to an appropriate precision.	
You plan to record physical conditions, such as temperature and pressure, if these affect the value of your dependent variable(s).	

9 Analysis

This criterion assesses the extent to which your report provides evidence that you have selected, recorded, processed and *interpreted* the data in ways that are relevant to the research question and can support a conclusion.

Mark	Descriptor
0	The student's report does not reach a standard described by the descriptors below.
1–2	The report includes **insufficient relevant** raw data to support a valid conclusion to the research question.
	Some **basic** data processing is carried out but is either too **inaccurate or too insufficient to lead to a valid** conclusion.
	The report shows evidence of little consideration of the impact of measurement uncertainty on the analysis.
	The processed data is incorrectly or insufficiently interpreted so that the conclusion is invalid or very incomplete.
3–4	The report includes relevant but incomplete quantitative and qualitative raw data that could support a simple or partially valid conclusion to the research question.
	Appropriate and sufficient data processing is carried out that could lead to a broadly valid conclusion but there are significant inaccuracies and inconsistencies in the processing.
	The report shows evidence of some consideration of the impact of measurement uncertainty on the analysis.
	The processed data is interpreted so that a broadly valid but incomplete or limited conclusion to the research question can be deduced.
5–6	The report includes sufficient relevant quantitative and qualitative raw data that could support a detailed and valid conclusion to the research question.
	Appropriate and sufficient data processing is carried out with **the accuracy** required to enable a conclusion to the research question to be drawn that is fully **consistent** with the experimental data.
	The report shows evidence of full and appropriate consideration of the impact of measurement uncertainty on the analysis.
	The processed data is correctly interpreted so that a completely valid and detailed conclusion to the research question can be deduced.

Table 9.1 Mark descriptors for the analysis criterion © IBO 2014

Recording and presenting raw and processed data

Tabulating data

Your individual investigation will involve recording qualitative data (observations) and quantitative data (measurements), often in tables. Raw data are measurements that are directly determined by recording analogue or digital readings.

In a table of data, the independent variable is put in the left-hand column, followed to the right by the dependent and processed variables. Typically the table will have the values of the independent variable listed in ascending or descending order. Units and the uncertainty should be in the column headings, not in the body of the table.

Values of the controlled variables can be written under the table (see Table 9.2).

Data consisting of very large or very small numbers is better shown in scientific notation (see Chapter 5).

> **Examiner guidance**
>
> You must not record raw data in imperial units, for example, in pounds, quarts, ounces, gallons, or feet and inches; SI units must be used. Times should be recorded in the SI base unit of seconds, not in minutes, and certainly not in a mixture of minutes and seconds.

Load/N (±0.1 N)	Reading at end of rule/cm (±0.005 cm)		Mean reading/cm (±0.05 cm)	Depression of cantilever/cm (±0.005 cm)	Time for 30 oscillations/s (±0.5 s)		Mean period of one oscillation, T/s (±0.5 s)	T^2/s² *
	Load increasing	Load decreasing			Trial 1	Trial 2		

Length of cantilever = 1000.000 cm (±0.005 cm)

* Here, the random error should be expressed as a percentage uncertainty

Table 9.2 Table of results for an investigation involving the depression of a cantilever by a load at the end

All raw data collected (with uncertainties and units) should be recorded, not just means of several experiments. If there is a relatively large amount of numerical raw data (for example, from a data logger), you may present one set of raw data for one trial and averaged data for the other trials.

If an experiment results in limited and unreliable data, you should consider the need to modify the experimental conditions to improve the data. This preliminary unsatisfactory data should be recorded in your IA report and commented on in your evaluation. The modifications that you make will be considered under 'personal engagement'.

■ Qualitative data

It is important to record relevant qualitative data during your investigation. These comments are especially relevant if they support your numerical data; for example, 'the two wires/magnets repelled/attracted each other' or 'condensation/ice formed on a cooled surface'.

They may provide evidence for random errors, for example:

- smudge dots on ticker tapes
- parallax problems in reading repeated measurements from a scale
- human reaction time in starting and stopping a timer
- random fluctuations in the read-out of an instrument
- blurring of an image in an optics experiment
- difficulties in judging when a moving ball passes a given point.

They may mention systematic errors, for example, ticker tape getting caught in the timer due to friction, or lack of control in variables, for example, if a spring was permanently deformed, a circuit breaker 'tripped' under certain conditions, or there was hysteresis in a magnetism experiment.

> **Examiner guidance**
>
> You should try to quantify observations listed above. For example, if you measured a voltage from an unstable, fluctuating power supply, you might write the following qualitative and quantitative comments: 'The voltage varied slightly with time, increasing and decreasing by several hundredths of a volt. Therefore, the values recorded have an uncertainty greater than the least significant digit of each measurement. The uncertainty was estimated to be ±0.05 V.'

Data processing

Complete and correct quantitative *processing* of the averaged raw data needs to be carried out.

In calculations, any intermediate results should have two extra digits beyond the last significant one, but the raw data and final results must be quoted with the correct number of significant figures (sf). Full calculations are not expected to be shown: selected examples will be sufficient.

The number of sf indicates the precision of measurements. If the times for ten oscillations of a simple pendulum are 24.2 s, 24.8 s and 24.6 s, and each reading has an uncertainty of the maximum deviation from the mean, which in this case is ±0.3 s, the average time for one oscillation = $\frac{24.2s + 24.8s + 24.6s}{30}$ = 2.453 333... s = 2.45 s (3 sf). The percentage uncertainty in the processed value is equal to the percentage uncertainty in the mean of the measured values. The final processed and raw data should both have 3 sf.

There are special rules for the precision with which to quote calculated logarithms and exponentials – see Chapter 5.

> **Examiner guidance**
>
> A measured value and its uncertainty can be expressed, for example, 12.5 cm ± 0.1 cm or (12.5 ± 0.1) cm. The following examples are incorrect: (12.45 ± 0.1) cm; (12.5 ± 0.02) cm and (12 ± 0.1) cm. The number of dp in the measurement and in the uncertainty must match in the raw data. See Chapter 5.

> **Expert tip**
>
> Are your measured values for the dependent variable in agreement with the prediction (perhaps based on a hypothesis)? Do the experimental data recorded fit the physical model? Are two measured values of a physical quantity the same?
>
> You cannot answer these questions without considering the uncertainties, that is the ranges, of your measurements. To make a judgment about whether two measured values X and Y are the same, you have to find the ranges in which these values lie. If the ranges $X \pm \Delta X$ and $Y \pm \Delta Y$ overlap, you can claim that the values X and Y are the same within your experimental uncertainty. If your percentage uncertainty is greater than your percentage difference from the accepted value, then your values are in agreement and you can claim your experiment is accurate (if not precise).

> **Expert tip**
>
> Collecting and presenting data is essential to identifying trends and patterns between an independent variable and dependent variables. Data processing and analysis become difficult or impossible if adequate data are not collected or if they are not properly recorded. You can use Excel, data-logging software and graphing software to process, analyse and present data.

Presenting data – graphing

Many physics investigations involve recording sets of repeated data for a pair of variables. The processed averaged data are then displayed, often as a line graph with data points showing error bars, ideally as a linear relationship between two variables. The equation of the line and the correlation coefficient may be shown.

Plotting pairs of data points that result in a curved graph also allows a general trend or relationship to be determined. In Figure 9.1a, as x increases, y decreases, but it is not clear whether the relationship is an inverse (Figure 9.1b), inverse square (Figure 9.1b) or exponential relationship (Figure 9.1b), or none of these.

Your graphs will have scatter due to random errors and it is easier to draw a straight line of best fit through linear plots than through plots that suggest a curve (a non-linear relationship). The gradient and intercept of the linear plot may also provide additional physical information. In order to plot a linear graph, linearization of the variables may be required (see Chapter 5); for example, a linear relationship may only result when a log graph is plotted.

> **Examiner guidance**
>
> Quoting the equation for the graph does not in itself answer a research question. You must also explain what is shown by the graph in terms of the physical properties under study: the mathematical results must be given physical meaning.

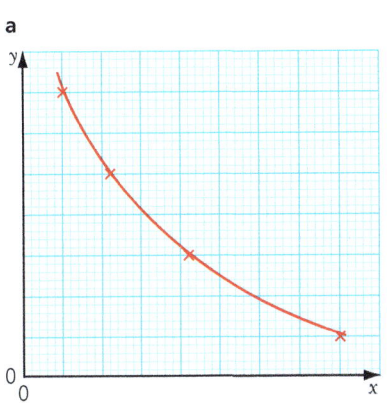 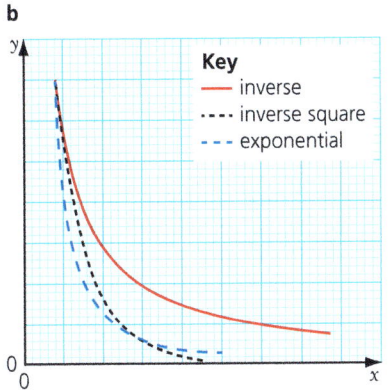

Figure 9.1 a As x increases, y decreases. **b** An inverse relationship ($y \propto 1/x$), an inverse square relationship ($y \propto 1/x^2$) and an exponential relationship ($y \propto e^{-x}$)

Worked example

You can sometimes check whether experimental data fit with a particular mathematical relationship without drawing a graph.

The data in Table 9.3 were collected for the tension, T, and the corresponding frequency, f, of vibrations on a guitar string.

Determine, without drawing a graph, whether the data are consistent with the expected relationship $f \propto \sqrt{T}$.

Tension in guitar string, T/N	115	150	175
Frequency, f/Hz	220	250	270

Table 9.3 Data from an investigation with a sonometer to determine the relationship between the tension in a guitar string and its frequency

Dividing each of the three values of the frequency, f, by the square root of the tension, T, gives a value of 20.4. This approximately constant value suggests that the data do show the expected relationship

> **Expert tip**
>
> If there is a large amount of data, you need only draw (or have software generate) error bars for the smallest value datum point, the largest value datum point, and a few data points between these extremes.

> **Expert tip**
>
> Never connect the data points dot-to-dot (Figure 9.2). Your data are meant to reveal a physical relationship, often described by a mathematical function, and this will not give a zigzag graph line.
>
>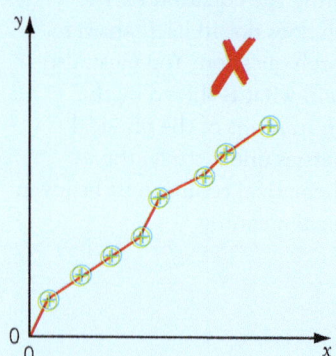
>
> **Figure 9.2** Incorrect plotting of a line graph

> **Common mistake**
>
> Students sometimes reduce a graph in size to ensure their report is within the page limit. You should not reduce a graph to such a size that it becomes uninformative.

Impact of uncertainty

You will need to carry out **error propagation** on all the measurements that you use in calculations. Different rules are used to calculate the final uncertainty from different types of mathematical processing: adding and subtracting values, multiplying and dividing values, raising to a power, multiplying or dividing by a constant, and applying trigonometric and logarithmic functions. All of these rules are detailed in Chapter 5. Errors in the gradients of linear graphs should also be estimated.

> **Key definition**
>
> **Error propagation** – A calculus derived statistical calculation designed to combine uncertainties from multiple variables, in order to provide an accurate measurement of uncertainty.

> **Worked example**
>
> The following data are from a simple pendulum experiment. The pendulum length and period were measured as $l = (2.53 \pm 0.02)$ m and $T = (3.2 \pm 0.1)$ s.
>
> The relationship being used is:
>
> $$T = 2\pi \sqrt{\frac{l}{g}}$$
>
> Determine g and its uncertainty using the data.
>
> Rearranging the equation:
>
> $$g = \frac{4\pi^2 l}{T^2} = \frac{4 \times \pi \times \pi \times 2.53\,\text{m}}{(3.2\,\text{s})^2} = 9.753\,944\,975\ \text{m s}^{-2}$$
>
> Calculating percentage uncertainties:
>
> $$\Delta l\% = \frac{0.02\,\text{m}}{2.53\,\text{m}} \times 100 = 0.7905\%$$
>
> $$\Delta T\% = \frac{0.1\,\text{s}}{3.2\,\text{s}} \times 100 = 3.125\%$$
>
> $$\Delta T^2\% = 2 \times 3.125\% = 6.250\%$$
>
> The percentage uncertainty in g can be taken as 7.0405% which can be converted back to an absolute error.
>
> $g = 9.753\,944\,975$ m s^{-2} $\pm\ 0.686\,73$ m s^{-2}
>
> $g = (9.8 \pm 0.7)$ m s^{-2}
>
> The uncertainty in g means that it should be expressed to 2 sf. (It is also acceptable that if one of the random percentage uncertainties is less than a quarter of the other, then the smaller percentage (in this case $\Delta l\%$) can be ignored, but this must be noted in your report).

> **Worked example**
>
> Find the absolute error in the reciprocal of the following volume measurement.
>
> $V = (32.5 \pm 0.4)$ cm^3
>
> $y = \dfrac{1}{V} = \dfrac{1}{32.5}$ cm^3 = 0.030 769 23 cm^{-3}
>
> % uncertainty in y = % uncertainty in $V = \dfrac{0.4}{32.5} \times 100$
>
> = 1.230 769 231% = 1.231%
>
> $\Delta y = \pm 0.000\ 378\ 698\ 224\ 9$ cm^{-3} = ± 0.000379 cm^{-3}
>
> therefore $y \pm \Delta y = (3.08 \pm 0.04) \times 10^{-2}$ cm^{-3}

> **Examiner guidance**
>
> It would be appropriate to present one full worked example of error propagation and not show for other similar calculations.

Interpreting processed data

Interpretation is the process of making physical and mathematical sense of the data that will lead to your **conclusion**. You should describe trends in graphs with reference to maxima (peaks) and minima (troughs), gradients (zero, positive or negative, increasing/decreasing) and intercepts (on the x- or y-axis). You should identify the mathematical relationship shown (such as linear, directly proportional, inversely proportional, negative or positive exponential). Where possible and appropriate, linearize non-linear relationships.

> **Expert tip**
>
> The plotted quantities have a *linear* relationship if the graph is a straight line. They are *proportional* only if the linear graph goes through the origin.

Rejecting data

In general, there are two different types of experimental data recorded in a physics lab:

- A series of measurements taken with an independent variable changed for each data point. An example is the calibration of a thermocouple, in which the output voltage is measured when the thermocouple is at a number of different temperatures.
- Repeated measurements are made of the same physical quantity, with all variables held as constant as experimentally possible. An example is measuring the time for a simple pendulum to undergo 20 oscillations and repeating the measurement five times.

How the rejection of measurements is decided on is different for each of these.

For a series of measurements, when one of the data points is an outlier and is apparently anomalous, the natural reaction is to reject it. But, as a 'rule of thumb', unless there is a physical explanation of why the suspect value is unreliable and it is no more than two standard deviations away from the expected value, it should probably be retained.

Rejecting a measurement is usually justified only if both of the following are true:

- A definite physical explanation is found of why the outlier value is unreliable.
- The outlier value is more than three error bars' length away from the expected (interpolated/extrapolated) value, and no other measurement appears unreliable.

Consider now the example of measuring the time for a pendulum to undergo 20 oscillations. Assume that you repeat the measurement five times; four of these trials are within 0.3 s of each other, but the fifth trial differs from the others by 1.6 seconds (which is more than two standard deviations away from the mean of the apparently reliable values). In this case, you may be justified in rejecting the fifth value.

> **Key definition**
>
> **Conclusion** – A reasoned interpretation based on experimental data.

> **Examiner guidance**
>
> *All* of your raw data should be recorded and any rejection must be justified, ideally statistically, and with a suggested physical reason for its relatively high or low value compared to other data.

> **RESOURCES**
>
> More information on detecting a single outlier can be found at:
>
> https://tinyurl.com/y9ltvtsg

Validity

Validity relates to the methodology of your investigation and how appropriate it is in addressing the research question. Does it test and measure what it was planned and meant to test and measure?

The following aspects of the experiment can contribute to its validity:

- the apparatus, materials and instruments
- the experimental method
- the analysis of the results.

The investigation is likely to be establishing a causal relationship between cause (independent variable) and effect (dependent variable): how changing X affects Y. You must systematically change X and measure the changes in Y, ensuring a fair test. If you allow other changes at the same time, then you cannot make a valid conclusion about how X affects Y, since Y may be affected by the changes in other variables as well.

The method (including the analysis) may contain some assumptions that need to be justified, for example, an equation used in a mathematical model may be an approximation. In an experiment with transformers, for example, the transformer equation $\frac{N_p}{N_s} = \frac{\varepsilon_p}{\varepsilon_s}$ is likely to be used. This equation, however, assumes there is no magnetic flux leakage, so a ferromagnetic core must be used. If it is not, then this assumption is not justified and the experiment will be invalid.

If your investigation is invalid then the experimental results have no meaning. The equipment, the methodology or the analysis was not appropriate for resolving the research question.

> **Key definition**
>
> **Validity (of methodology)** – Suitability of the investigative methodology to answer the research question.

Analysis criterion checklist

■ Recording raw data

Descriptor	Complete
You neatly record all raw data: qualitative data (observations) and quantitative (numerical) data necessary to support a conclusion to the research question.	
You record an appropriate number of trials for both the independent and the dependent variables.	
You clearly record single measurements that are supplemental (related to controlled variables, not the independent or dependent variable), with appropriate units.	
You present all raw numerical data clearly and correctly in tabulated manner (for example, independent variable on far left, then dependent variable followed by processed variable(s)). SI units are typically used.	
The headings in your data tables have labels, units and uncertainties (absolute errors) once in the heading.	
You consistently record quantitative data, taking into account the absolute uncertainty/error (correct number of decimal places).	
If using scientific notation, you quote the value and the error with the same exponent.	
You describe how the absolute uncertainties in the measurements were obtained, for example: half of the smallest division of the scale, smallest digital number displayed or manufacturer's tolerance.	
You discuss reaction time for manual timings.	
You clearly indicate and highlight any anomalous data (and statistically justify its exclusion, if appropriate).	
Your processed data resolve the research question.	

Processing raw data

Descriptor	Complete
You use the appropriate formulae and annotated mathematical equations to carry out calculations to correctly process raw data (for example, averages, reciprocals, trigonometric functions or logarithms to base 10 or e).	
You report results of calculations according to the rules of significant figures.	
At least one complete sample calculation is included for every individual type of calculation used during processing.	
Sample calculations are located immediately after the table in which the data are reported.	
You convert the appropriate averaged data into the correct graphical form to show the relationship between the independent and dependent variables.	
Each graph contains point symbols, but no connecting lines.	
Where relevant you select appropriate processed average data to produce a straight line graph (if the data are continuous) with a line of best fit (graphically or computer-generated).	
If the initial graph is not linear, you give sample calculations and a data table showing the mathematical manipulation done to one of the axes of data.	
You extract a relevant quantity (or quantities) from the graph, for example, the gradient of a straight line, gradients along a curve, intercept (or sometimes area), or perform interpolation or extrapolation.	
You process repeated data – finding the mean and using the variation between values to assign an appropriate uncertainty.	

Presenting data

Descriptor	Complete
You use scientific conventions in tables of processed data.	
You use accepted conventions for graphs, for example: title, suitable graph size (large), appropriate range and scale, labelling and units.	
Your graphs usually have the independent variable on the x-axis and the dependent variable on the y-axis.	
You draw the line or curve of best fit correctly and clearly indicate the trend shown.	
Your graphs include the equation and the coefficient of determination (R^2), where appropriate.	
You include error bars on line graphs, where appropriate and possible.	
If the error bars are too small to be visible, you include a statement explaining this under the graph.	

Impact of uncertainity

Descriptor	Complete
You report appropriate absolute uncertainties for each measurement associated with the independent and dependent variables.	
You report appropriate absolute uncertainties clearly for each of the controlled variables.	
You convert absolute uncertainties to percentage errors and propagate uncertainties appropriately and consistently (via selected sample calculations).	
You present calculations and error propagation in a clear, organized and separate manner.	
You present final figures with the number of decimal places consistent with the propagated uncertainty.	
You have included maximum and minimum gradients on the final linear graph.	
Maximum and minimum gradients are appropriate for the reported uncertainties/ error bars used.	

Interpreting data

Descriptor	Complete
You report the gradients for each of the three lines (best-fit, maximum and minimum) with an appropriate number of sf and correct units.	
You report the final gradient clearly with appropriate uncertainty, sf and units.	
If outliers were present in the data, you include an explanatory statement to outline why these values were determined to be anomalous, and what was done with them during data processing.	
You interpret the processed data correctly, for example, the correct relationship between the independent and the processed dependent variable is deduced from a graph, with quoted data to support it.	

10 Evaluation

This criterion assesses the extent to which your report provides evidence of evaluation of the investigation and the results, with regard to the research question and the accepted scientific context.

Mark	Descriptor
0	The student's report does not reach a standard described by the descriptors below.
1–2	A conclusion is outlined which is not relevant to the research question or is not supported by the data presented.
	The conclusion makes superficial comparison to the accepted scientific context.
	Strengths and weaknesses of the investigation, such as limitations of the data and sources of error, are outlined but are restricted to an account of the practical or procedural issues faced.
	The student has outlined very few realistic and relevant suggestions for the improvement and extension of the investigation.
3–4	A conclusion is described which is relevant to the research question and supported by the data presented.
	A conclusion is described which makes some relevant comparison to the accepted scientific context.
	Strengths and weaknesses of the investigation, such as limitations of the data and sources of error, are described and provide evidence of some awareness of the methodological issues involved in establishing the conclusion.
	The student has described some realistic and relevant suggestions for the improvement and extension of the investigation.
5–6	A conclusion is described and justified which is relevant to the research question and supported by the data presented.
	A conclusion is correctly described and justified through relevant comparison to the accepted scientific context.
	Strengths and weaknesses of the investigation, such as limitations of the data and sources of error, are discussed and provide evidence of a clear understanding of the methodological issues involved in establishing the conclusion.
	The student has discussed realistic and relevant suggestions for the improvement and extension of the investigation.

Table 10.1 Mark descriptors for the evaluation criterion © IBO 2014

Common mistake

Evaluation is often the weakest part of an IA report. It is a difficult skill to learn and apply, and sometimes candidates finish off the report in a hurry, leading to an inadequate evaluation.

Concluding

Drawing justified conclusions is a skill that involves analysing the results of an investigation, stating and explaining what they show. This is done by analysing and displaying processed data and using the data to focus on answering the research question. The conclusions should be evaluative (based on the data) rather than descriptive (stated in general terms) and make use of physical and mathematical ideas from your background knowledge.

Physics investigations may allow you to discover or confirm mathematical relationships between variables. A conclusion often explains how the independent variable causally affects the dependent variable. If you find there is no relationship between variables, that is as important a conclusion as establishing a relationship. If a hypothesis has been proposed then you should conclude whether the data support it or do not support it.

You must establish whether the trends or relationships (shown by your graphs) are consistent with the accepted physical theory and physics literature (which you should reference). If you are unable to do this, then you need to demonstrate that your raw data are reliable and that your data processing is appropriate.

> **Worked example**
>
> The line of best fit of the voltage–current graph clearly lies within the range of all the error bars and passes nearly through the origin, thus establishing a linear and proportional relationship. From the gradient of the graph the resistance, $R = 1.1$ kΩ. The minimum and maximum gradients show that the resistance is $R \pm \Delta R = (1.1 \pm 0.1)$ kΩ. For the given range of voltage and current, a proportional relationship holds and obeys Ohm's law: the current through the conductor between two points is directly proportional to the potential difference across the two points.

For some investigations you may have determined the value of a physical constant (for example, acceleration due to gravity) or property (for example, resistivity). You should compare your experimental result with the literature value. The measured and accepted values are in agreement if the accepted value lies within the range defined by the measured value ± its random uncertainty.

To report the accuracy of your result, calculate the percentage error (that is the percentage discrepancy between the two values). Take the absolute value of the difference between your value and the literature value, divide this difference by the accepted value and multiply by 100. Taking x to be your measured value and y to be the literature value:

$$\text{percentage error} = \left| \frac{x - y}{y} \right| \times 100$$

> **Worked example**
>
> Calculate the percentage error in the experimental measurement of g with a value of 9.5 m s^{-2}.
>
> $$\text{percentage error} = \left| \frac{9.5 - 9.8}{9.8} \right| \times 100 = 3.06\%$$
>
> The measured value with its experimental uncertainty was 9.5 ± 0.4 m s^{-2}. Does this agree with the accepted value?
>
> Yes, the two values agree, within the uncertainties of the experiment.
> The accepted or literature value is 9.8 m s^{-2}.
> $9.8 - 9.5 = 0.3$ which is less than 0.4.

Compare the discrepancy between the experimental and the literature value with the total random error (the final propagated uncertainty in your data) and comment on it. If the discrepancy is greater than the random error, then random error (which is always present) alone cannot explain the difference between your value and the literature value. This indicates the presence of systematic error in your experimental method.

Examiner guidance

You should discuss systematic errors when there is clear evidence for them, and be aware they might have cancelled each other out. The direction of a systematic error on the measured value should be included, that is whether it made the experimental value higher or lower than expected.

Note that human (or personal) errors are 'mistakes' (see Figure 10.1) and can be avoided with careful technique. Examples include: misreading a scale, making a calculation mistake, incorrectly plotting a reading on a graph, handling equipment poorly (for example, spilling a liquid), or being prejudiced in favour of earlier over later data readings.

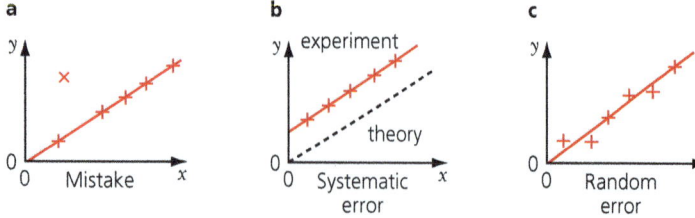

Figure 10.1 Graphs showing a mistake, a systematic error ('bias') and random errors ('scatter') in an experimentally investigated linear relationship

Examiner guidance

Beware of claiming in your IA report that 'personal' or 'human' errors limited the accuracy of an experimental result. This is to admit to incompetence.

> **Examiner guidance**
>
> Remember that hypotheses enable the design of investigations so that predictions may be tested, and either supported or falsified. Do not overstate your findings. Always be tentative in your conclusions. For example, 'the graph *suggests* a likely relationship between variables x and y'. If your data appear to support your hypothesis, re-state the hypothesis in the context of your data and methodology.

> **Expert tip**
>
> A hypothesis *suggests* the true nature of the relationship being tested. It remains valid even when the results of the investigation contradict it. For example, a hypothesis might be: 'If the period of a simple pendulum is related to its length, then the longer the pendulum the shorter the period.' When experimental results show the opposite to be true, the hypothesis is still *valid*, because it has allowed the investigation to be focused.

In summary:

- base all conclusions in your report on your actual experimental results
- explain the outcome of the practical investigation (and any simulation) and the implications of your results
- examine the outcome in the context of the research question – do the data fit the model (if implied in the research question) to within the uncertainties present in the measurements?

Evaluating strengths and weaknesses of the investigation

Your methodology must be evaluated thoroughly. You need to include the strengths as well as the weaknesses of the investigation. For example, a strength could be a small range (or small standard deviation) in data for the dependent variable, indicating that repeated measurements were precise. A weakness may be poor control or lack of control of some variables. Do not restrict weaknesses to details of the practical, but include other aspects as well, for example, lack of secondary data or literature values for comparison.

■ Limitations of the data and sources of error

In this part of the report you should discuss the uncertainties (percentage error) in your measured dependent and independent variables. The impact of the limitations on the conclusion must be discussed.

You should refer to random and systematic errors (Table 10.2). Make sure you have calculated the total (propagated) percentage random error. Where possible, give the size and the effect of probable systematic errors.

Systematic errors	Random errors
These produce a constant **bias** in the results.	These consist of varying fluctuations in the system.
Averaging repeated readings will not reduce them.	They can be both higher and lower than the true value.
They can only be reduced by setting up the apparatus carefully, accurate calibration of the instrumentation and cross-comparison with known values.	Their effect can be reduced by averaging repeated readings.
It is difficult to study them statistically.	They can be studied statistically.

Table 10.2 Comparison of systematic and random errors

> **Key definition**
>
> **Bias** – When results are shifted in a particular direction by a systematic error.

> **Expert tip**
>
> In a single measurement it is not possible to isolate the systematic and random errors. They combine together to give a final error as illustrated in Figure 10.2. The bell curve shows the random error and the shift to the right shows the effect of systematic error.
>
>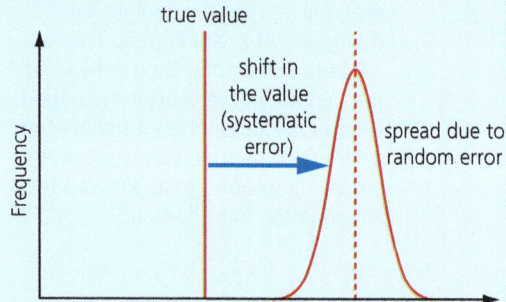
>
> **Figure 10.2** The error in every measurement is a combination of systematic and random errors

Examine the error propagation – how the random errors have been propagated to obtain the final uncertainty. You may identify one of the quantities measured as a critical measurement, that is, with the largest percentage uncertainty. Suggest what approach could be used to measure that quantity with a higher resolution, that is more precisely.

Focus on these major sources of error and how they could be reduced. For example, in a calorimetry experiment where there is a 50% systematic error due to heat loss, this is the major error to be addressed/improved/minimized by adjusting the methodology.

> **Examiner guidance**
>
> Ideally, you would identify a critical measurement during your preliminary work and modify your methodology to take account of this.

> **Examiner guidance**
>
> There is little to be gained by discussing small sources of random error such as uncertainties in masses or volumes measured in such an experiment. If you just list every possible source of error without any consideration of the relative magnitude of each, you have not effectively evaluated your investigation.

> **Worked example**
>
> Experimental data from investigating the relationship between pressure and volume of a fixed mass of air (at constant temperature) were graphed as shown in Figure 10.3. The dotted line indicates the expected relationship from Boyle's law (assuming ideal behaviour). The data were obtained from a syringe connected to a Bourdon pressure gauge in a laboratory.
>
> Discuss the results shown and possible systematic errors.
>
>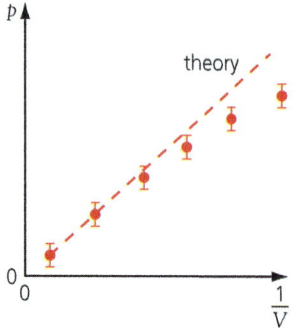
>
> **Figure 10.3** Plot of pressure versus reciprocal of volume (for a fixed mass of air at constant temperature)
>
> It can be observed that as the volume decreases and the pressure gets larger, the gradient of the graph decreases slightly. Likely sources of systematic error in this experiment include: the volume of air in the rubber tubing is not measured (known as the dead space); there is leakage of air; the temperature of air inside the syringe is not constant.

In summary: comment on the overall validity of the method, with reference for example, to the range (of repeated measurement of the dependent variable), and discuss the possible use of an alternative methodology.

■ Assumptions

You must outline any assumptions present in your methodology and calculations. These assumptions may be practical, mathematical or physical in nature. For example:

- In an investigation involving the bending of a beam, the theory of simple bending may make the following assumptions: Hooke's law is valid for both tensile and compressive stresses; there are no shearing stresses over any section of the beam when it is bent; there is no change in the cross section of the beam on bending.
- In an investigation involving a sonometer, air resistance would not be included in the simple physical model of standing waves on a string.
- When studying and measuring the magnetic field of a small magnet you cannot ignore the effect of the Earth's magnetic field.
- The dielectric material inside capacitors is not ideal. It involves losses that are frequency and voltage dependent, appearing as ohmic losses.

Improvements and extensions

Your report needs to propose improvements to your investigation, and these must be realistic and specific. Suggest improvements and/or extensions to your methodology that are precise, focused and relevant to your investigation. They should be directly related to the weaknesses or limitations that you identified and should be feasible to be carried out in a school physics lab. Any suggested extensions should follow on from your research findings and, if performed, would enhance your understanding. A tabulated approach may be helpful as shown in Table 10.3.

Methodology limitation	Impact on results	Suggested improvement

Table 10.3 One possible tabular approach to proposing improvements

You should always ensure existing trials of the investigation are repeated a minimum of twice (or more if there is any evidence of anomalous data points). Repeated results should be averaged to reduce random uncertainty and hence increase accuracy. You should compare repeated results to consider their precision and hence how reliable they are.

> **Worked example**
>
> Using multiple slits is similar to performing repeats. Diffraction gratings are used to make very accurate measurements of the wavelength of light. In theory, they function much the same as Young's double-slit experiment (Chapter 4). However, a diffraction grating has many slits, rather than two, and the slits are very closely spaced. By using closely spaced slits, the light is diffracted to large angles and measurements can be made more accurately (with smaller random uncertainty). In spreading out the available light to large angles, brightness is lost; by using many slits (that is sources of light) brightness is preserved.

> **Expert tip**
>
> One of the best ways to obtain more precise measurements is to use a 'null' method instead of measuring a quantity directly. This involves using instruments to measure the *difference* between two similar quantities, one of which is known very accurately and is adjustable. The adjustable reference quantity is varied until the difference is reduced to zero. The two quantities are then 'balanced' and the size of the unknown quantity can be found by comparison with the reference sample. The use of a potentiometer and the Wheatstone bridge are examples of null methods. With this method, problems of source instability are eliminated, and the measuring instrument can be very sensitive and does not even need a scale.

> **RESOURCES**
>
> - https://opentextbc.ca/physicstestbook2/chapter/null-measurements/
> - https://www.electronics-tutorials.ws/blog/wheatstone-bridge.html

> **Expert tip**
>
> Another reason for repeating is that it allows you to identify an unreliable result that is the result of a mistake in carrying out the experiment. Be selective about the data you are going to average, including only consistent results.

Consider an experiment to investigate the capacitance of a capacitor. Selected limitations and improvements are summarized in Table 10.4.

Limitation	Improvement
The precision in the measurement of the resistance is relatively low.	Use a digital voltmeter that records to ±0.01 V. Take several values of pd/current and calculate the mean.
It is difficult to observe the stopwatch and ammeter at the same time to record data. The current reading lags the time measurement.	Use a data logger and retrieve the values later for analysis.
There is difficulty in reading a changing current.	Slow the investigation down by using a higher value resistor, or use a data logger.
There might be a systematic error in the stopwatch (for example running slow due to battery) or in the microammeter.	Use a recently calibrated stopwatch and microammeter.

Table 10.4 Limitations and improvements of an investigation into the discharge of a capacitor

Worked example

Here are some evaluative comments from an investigation into a factor that affects the resistance of a wire.

Unreliable results were recorded when investigating the two thinnest wires, nichrome 0.30 mm and copper 0.49 mm. The assumption had been made that the wires have a consistent diameter (within the experimental error of the micrometer (±0.01 mm)) throughout their length, but this is unlikely. Also, the thinner wires heat up more easily with current flow.

Waiting for the wire to cool after each ohmmeter measurement would have increased the accuracy of the resistance measurements.

Measuring the diameter of the wire at 5.00 cm intervals may have given a larger data set, improving the statistical significance of the calculations and hence graphs.

Wires of diameter 1.00 mm or larger should be used, as the results have shown that very thin wires do not give reliable results. Wires of known composition (pure metals and alloys such as nichrome) are best to use to allow comparison. Alloys of variable composition which are prone to corrosion are less suitable, for example steels.

A further investigation could use the same lengths of wires but deliberately introduce kinks in the wire to see the effect on resistance values.

RESOURCES

http://www.ucl.ac.uk/~zcapf71/FormalReport2.pdf

Using secondary data

Where appropriate, you should have described and commented on how secondary data support your primary data (see 'Concluding', page 115), and identify any areas of incompleteness.

Ideally, you should have a range of relevant secondary data collected from several reliable sources. You should assess the level of confidence that can be placed on the available data from these sources (see 'Assessing sources', page 127), and explain the reasons for making these judgments.

In your IA report you must reference (acknowledge) all sources of secondary data, and also include a full bibliography at the end of the report. See 'Referencing', page 127.

Expert tip

Note that:

- .ac and .edu domains are educational sites
- .gov domains are government sites
- .co and .com domains are commercial sites
- .org domains are used by non-profit organizations.

Evaluation criterion checklist

■ Concluding

Descriptor	Complete
You include a detailed and logical conclusion (for example, trends between independent and dependent variables), citing numerical values (in data tables and graphs) relevant to the research question.	
If any hypotheses are being tested, there is a statement of whether the data support or do not support the hypotheses.	
The level of support (strong, weak, none, or inconclusive) for the hypothesis/conclusion is identified, correct and justified.	
Your conclusion is supported by the raw and processed data (typically graphs) and the observations.	
You include a scientific conclusion for the results that is described, justified and compared to the relevant physical literature (if available). Any literature reference values must be referenced.	
You identify and comment on any anomalous data (outliers).	
You discuss the accuracy of the conclusion quantitatively and qualitatively, referencing data and comparing to known values as appropriate.	
Your conclusion is based solely on your primary data (experimental results) and uses tentative words such as 'indicate', 'suggest', 'appear to suggest' and 'support'; not 'prove'.	
Your conclusion is relevant to the research question and is valid within the limits of random uncertainties.	
You justify your results with reference to relevant physical laws and models, theories and principles.	
You discuss limitations to your results.	

■ Evaluating procedures: strengths and weaknesses

Descriptor	Complete
You describe assumptions that were made which have affected the accuracy of the results, for example, ideal behaviour in gases or mixtures of pure liquids, incompressibility of liquids (constant density), laminar flow in a liquid.	
You calculate the percentage error between the experimental and literature values (if available and relevant).	
You identify systematic errors and their directional effect (increase or decrease) on the experimental results and the conclusion.	
You compare the percentage uncertainties (errors) of each type of data and identify the major uncertainties (errors).	
You discuss any limitations of the method, for example, limited data range, lack of data around inflexion points, relatively poor precision of readings and limited instrument sensitivity.	
Obvious errors/limitations have not been overlooked.	
The identified errors/limitations are not overly simplistic, or negligible in magnitude.	

■ Improving and extending the investigation

Descriptor	Complete
You suggest appropriate modifications in the steps taken to improve the accuracy, precision and reliability of the results (by reducing random and systematic errors) or to improve control/monitoring of controlled variables.	
You suggest a reasonable alternative method or different instrumentation to obtain the same experimental data, more accurately.	
You discuss clearly how the suggested improvements or modifications focused on the errors/limitations would improve the reliability, precision and accuracy of the results.	
You propose realistic and relevant extensions to the investigation, for example, greater range of data, more data from around an inflection point, revised data processing/data presentation, choice of new independent variable.	
Your suggested improvements are focused on the existing research question.	
A new research question (extension) may be stated with a clear independent and dependent variable.	
The new research question is an extension from the conclusion and evaluation. A short explanation for the new research question is given to establish its importance and relevance.	

11 Communication

This criterion assesses whether you have presented and reported the investigation in a way that effectively communicates the focus, process and outcomes.

Mark	Descriptor
0	The student's report does not reach a standard described by the descriptors below.
1–2	The presentation of the investigation is unclear, making it difficult to understand the focus, process and outcomes. The report is not well structured and is unclear: the necessary information on focus, process and outcomes is missing or is presented in an incoherent or disorganized way. The understanding of the focus, process and outcomes of the investigation is obscured by the presence of inappropriate or irrelevant information. There are many errors in the use of subject specific terminology and conventions*.
3–4	The presentation of the investigation is clear. Any errors do not hamper understanding of the focus, process and outcomes. The report is well structured and clear: the necessary information on focus, process and outcomes is present and presented in a coherent way. The report is relevant and concise thereby facilitating a ready understanding of the focus, process and outcomes of the investigation. The use of subject specific terminology and conventions is appropriate and correct. Any errors do not hamper understanding.

* For example, incorrect/missing labelling of graphs, tables, images; use of units, decimal places. For issues of referencing and citations refer to the academic honesty section.

Table 11.1 Mark descriptors for the communication criterion © IBO 2014

Structure and clarity

The presentation of your IA report must be coherent and relevant to the focus (the research question), the process (methodology) and the outcomes (results and conclusion). It should be word processed with numbered pages and resemble a published scientific paper. There should be sections with headings and sub-headings to give structure (see 'IA report format', page 125). Diagrams and digital images should be used to enhance understanding. There should be a logical sequence, allowing the IB examiner to understand your thought processes throughout the report.

Your methodology should be detailed enough for the experiments to be reproducible, but simplistic, well known and assumed aspects need not be described.

It is best to use the past tense passive voice, for example, 'the potentiometer was balanced'.

It is best to avoid contractions, for example, it's, hasn't or didn't, in formal scientific writing. Use 'it is', 'has not', etc. instead.

> **Expert tip**
>
> When referring to the logarithm function or the sine function in the text of your report, refer to them as logarithm and sine, not log and sin, for example, 'taking the logarithm of $N(t)$', not 'taking the log of $N(t)$'.

Relevance and conciseness

Your report must be relevant to the research question. The development of your ideas and thoughts should be easy to follow from the beginning to the end of your report. It should be concise, with no unnecessary, irrelevant or repetitive information.

> **Expert tip**
>
> Be careful about writing 'calculated' when you mean 'measured', for example, 'calculated the resistance', which suggests that you may have obtained it theoretically from the resistivity, length and cross section of the wire.

> **Examiner guidance**
>
> Take care over the length of your report. It must be 12 pages maximum.
>
> It is advisable not to include an appendix, because this will reduce your available page count for the main report, but it will not be automatically marked down provided the report is relevant and concise.
>
> Do not include excessive quantities of raw data from a data logger.

Full calculations for processing all of the data and propagation of all the errors are not expected – selected examples are sufficient. This will free up more space for your conclusion and evaluation.

Do not overexplain the use of computers, calculators or software; they are tools. Focus on the physics.

Terminology and conventions

It is important that you use correct physics terminology, scientific conventions (for example, letters in mathematical equations are italicized; numbers should never be italicized) and units. You must label tables and graphs suitably, include an appropriate number of significant figures in your data and include random uncertainties.

> **Common mistake**
>
> Do not use a whole page for the title or the contents.
>
> Do not present blank data tables at the end of the method section.

> **Examiner guidance**
>
> Note these differences.
>
> | Hyphen | - | Minus sign | – |
> |---|---|---|---|
> | Letter x | x | Multiplication sign | × |
> | Superscript letter o | º | Degree sign | ° |
>
> Use the Insert Symbol function in Word to insert special symbols.

> **Examiner guidance**
>
> Spaces in units are important to distinguish between, for example, metre per second (m s^{-1}) and millisecond (ms).

In general, use SI units in your report unless non-SI units are appropriate in your topic. There should be a space between the numerical value and the unit, and between each unit. The only common exception to this rule is angular degrees (°), in which case no space should be included.

The negative power notation (m s^{-1}) is the IBO preference to the slash notation (m/s).

The abbreviations of units should never be in italics. But letters representing variables in equations, including Greek letters, should be in italics; for example, the resistivity of material of a wire is:

$$\rho = \frac{RA}{l}$$

Vectors can be written with an arrow, for example, $\vec{A}$.

Letters indicating a point on a graph or diagram are not considered as variables and hence should not be written in any special way.

When referring to graphs, do not use the generic x and y to refer to the horizontal and vertical axes. You used the graph to show the relation between two physical quantities, which will have a variable to represent them; use those variables instead.

Be careful with formatting physical quantities – make sure they are in the same format in your text and in your equations and graphs. For example, stick to V_o or V_{output}; do not use a mixture.

Equations may be embedded in the text of your report and formatted using the Equation Editor tool in Microsoft Word© (Figure 11.1). Equations can be numbered so that you can refer back to them later in the report. Where appropriate, the derivation of the equations should be included.

> **Expert tip**
>
> Do not use the letter w in place of the Greek letter omega, ω.

Figure 11.1 Equation Editor tool bar in Microsoft Word©

All tables, graphs and illustrations should be introduced by a sentence of explanation and have an explanatory descriptive label (caption). To improve the flow of the report, these should be positioned in the text so as to maximize the use of available space, ideally close to the location in the text where they are referred to. You must reference the source of any images that you include taken from the internet.

In circuit diagrams, all lines should be straight and there must be no gaps in the circuit. A gap may be interpreted as a disconnection. Wires and components are aligned horizontally or vertically, unless there is a good reason to do otherwise. Figure 11.2 shows the standard electric circuit symbols used in the IB Diploma Programme physics course. You may need to use additional symbols, such as for logic gates, an inductor, a microphone, an op-amp or transistors.

> **Common mistake**
>
> Some commonly misspelled or misused words are:
>
> accommodation, dependence, dependent, in principle, principal axis, (computer) program, (other type of) programme, physical phenomenon (singular), physical phenomena (plural), criterion (singular), criteria (plural), datum (singular), data (plural).

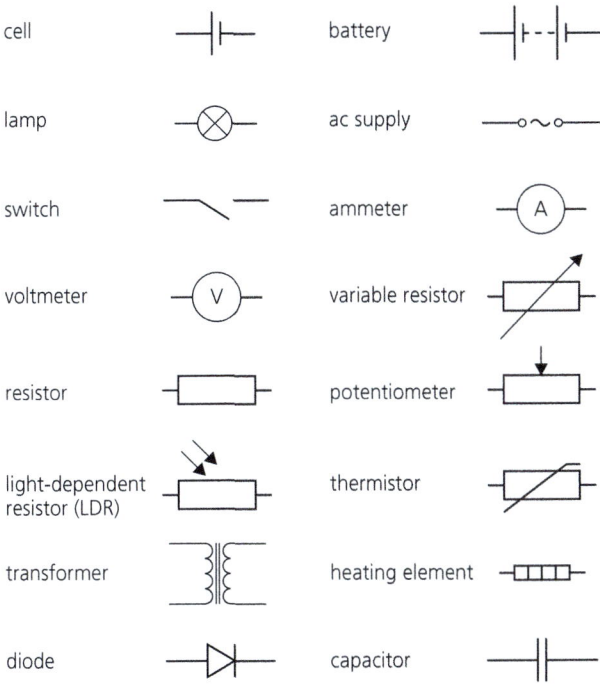

Figure 11.2 Circuit symbols

Make sure that the components are connected in the correct order and are labelled. Wires that connect should be indicated by a heavy black dot; wires crossing but not connecting have no dot.

IA report format

There is no particular structure that you must follow, but it should resemble a research paper (without the abstract). If your school does not suggest a format, then you could use the following headings:

- General title or aim
- Background information (physical theory, mathematical model including equations) and hypothesis (if appropriate)
- Personal engagement statement (if appropriate)
- Research question
- Risk assessment
- Planning and preliminary work
- Variables
- Method
- Raw data
- Processed data (including graphs)
- Error propagation
- Conclusion
- Evaluation:
    - Random and systematic errors
    - Limitations, weaknesses and improvements

- ☐ Comparison with secondary sources
- ☐ Further investigations
- ■ Bibliography of references.

Use Figure 11.3 to check that you have covered all aspects of each assessment criterion.

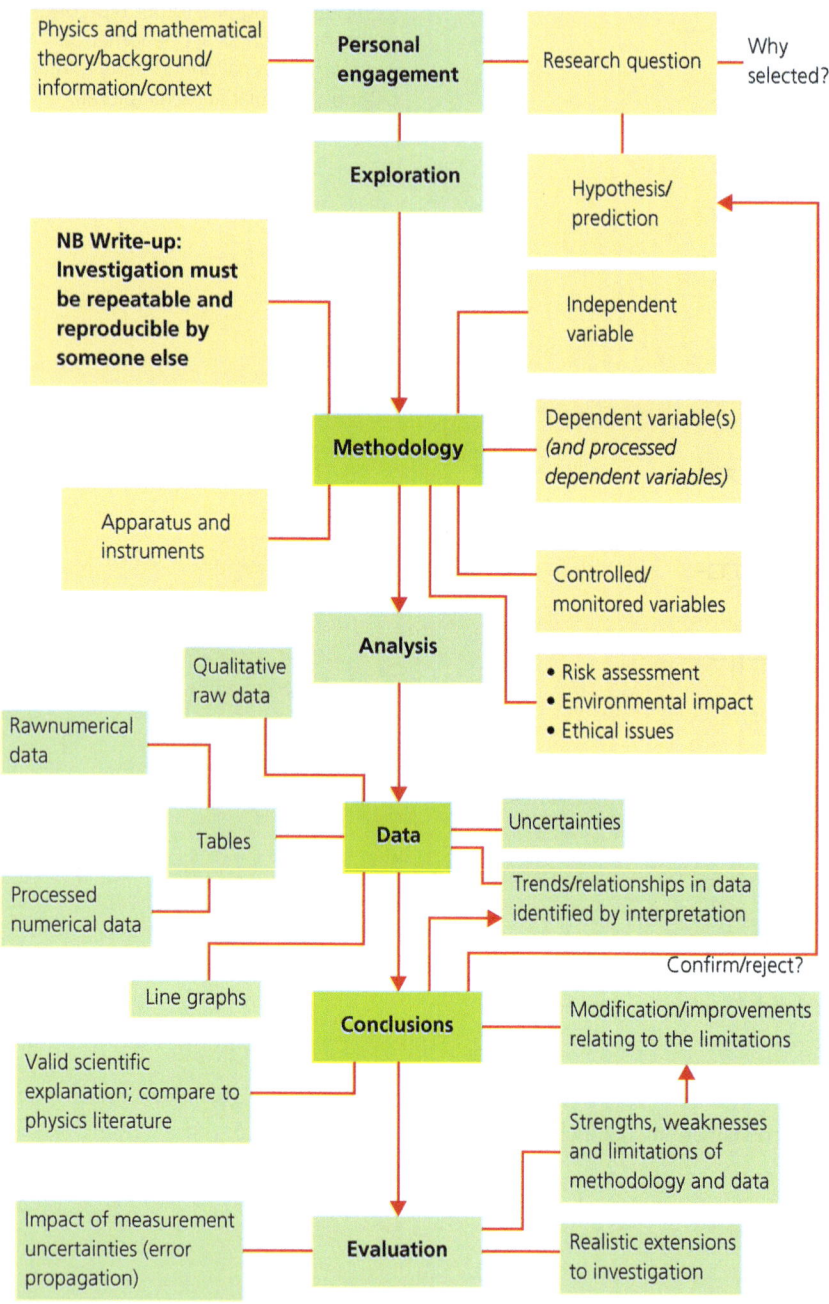

Figure 11.3 Physics internal assessment map

RESOURCES

https://ibpublishing.ibo.org/server2/rest/app/tsm.xql?doc=d_4_physi_tsm_1408_1_e&part=8&chapter=1

Examiner guidance

Communication, like personal engagement, is assessed holistically, meaning the entire report is taken into consideration.

Referencing

Referencing is a standardized method of acknowledging the sources of information you have consulted. Any words, paragraphs, quotations, figures, images, tables, theories, scientific ideas or facts originating from another source and used in your IA report must be referenced. The reasons for this are:

- to avoid plagiarism (presenting someone else's work as your own)
- so that your teacher can verify quotations
- so that your teacher can follow up on your thinking by consulting the source you accessed.

Your school or department is likely to have a referencing style that you must adopt. Familiarize yourself with the format and terms that your school or IB Physics teacher expects you to use. Be consistent.

A reference may consist of two parts: the *in-text citation*, which provides brief identifying information within the text, and the *reference list*, a list of sources that provides full bibliographic information. Here are some examples of entries in a reference list.

Book:

Allum, J. and Talbot, C., *Physics for the IB Diploma* (2nd edn), Hodder Education, 2014.

Journal article:

Stafford, O., 'Experimenting with a 'pipe' whistle,' *Phys. Teach.* 50, 229–230 (April 2012).

Website:

Walding, R., http://seniorphysics.com/physics/eei.html (accessed 29 May 2018).

■ Assessing sources

Before you decide to use some source material, ask yourself these questions:

- Can you identify the author's name?
- Can you determine what qualifications or titles he/she has?
- Do you know who employs the author, such as a university or company?
- Is this a primary source (original research paper) or secondary source (review article)?
- Is the content original or derived from other sources?

■ Evaluating information

It is important that you check the validity of the sources you are using. Do not assume that information is correct. Apply the following checks:

- Have you checked a range of sources for this information?
- Is the source information supported by relevant literature citations?
- Is the age of the source likely to be important regarding the scientific accuracy of the information?
- Is the information scientific fact or opinion/speculation?
- Have the data been analysed (if appropriate) using relevant statistics?
- Are the data in graphs displayed fairly with error bars?

> **Expert tip**
>
> Scientific papers submitted to peer-reviewed journals, such as *Nature*, are carefully scrutinized by experts in the field ('peers'). Information from such sources can be trusted as being scientifically valid.

Academic honesty

Academic honesty means having values and skills that promote personal integrity and good practice in teaching, learning and assessment. Learners having academic honesty and integrity is consistent with the IB learner profile (see 'Studying IB Physics' at the start of this book).

You need to ensure that the IA report that you submit is your own work. At the end of the course you will need to sign a declaration. Your IB Physics teacher will check the authenticity of your work by:

- discussing it with you
- asking you to explain the method and to summarize the results and analysis
- possibly asking you to repeat part of the investigation
- using dedicated text-checking software and websites to check for plagiarism.

> **RESOURCES**
>
> https://www.ibo.org/globalassets/digital-tookit/brochures/effective-citing-and-referencing-en.pdf

Plagiarism

Plagiarism is defined (by the IBO) as the representation, intentionally or unwittingly, of the ideas, words or work of another person without acknowledgment. Examples of plagiarism include:

- copying the work of another IB Physics student (past or present) or getting someone else to write your report (or part of it) and passing it off as your own work
- copying text or images from a source (for example, a book, journal article or website) and using it without acknowledgement
- quoting others' words without indicating who wrote them or said them
- copying scientific ideas and concepts from a source without acknowledgement, even if you paraphrase them.

> **Expert tip**
>
> To avoid plagiarism:
> - make sure the work (results and theory) you present in your IA report is always your own
> - never 'copy and paste' from websites or files downloaded from the internet
> - include appropriate references in your report
> - show clearly where you are quoting directly from a source.

Communication criterion checklist

■ Structure of report

Descriptor	Complete
Your report is well structured, coherent and clear, following the style and conventions of a scientific research or review paper.	
Your report is split into appropriate sections (with headings and titles): focus, process and outcomes, the IB criteria or other headings.	

■ Relevance and conciseness

Descriptor	Complete
Your report includes only relevant information and is concise (6 to 12 pages).	
Your report does not contain any mistakes, contradictions, false scientific statements or false assertions.	

■ Subject-specific terminology and conventions

Descriptor	Complete
Your graphs, tables and images are fully titled and referenced and presented according to accepted conventions in scientific papers.	
Your mathematical equations are in italics and clearly explained and justified/derived (where appropriate) perhaps with appropriate units.	
You have followed the rules relating to significant figures (sf). The number of decimal places (dp) is correct in data tables and calculations.	
You have defined non-syllabus terms and abbreviations.	
Your report includes a cross-referenced bibliography with in-text referencing according to a particular referencing style.	

Glossary

Absolute uncertainty (absolute error) – The uncertainty or error in a measurement that is expressed in physical units.

Accuracy – How close a measured value is to the true or theoretical value.

Analysis – Recognizing and commenting on trends in raw and processed data and stating valid conclusions.

Anomalous data – Data with unexpected values (beyond the limits expected by uncertainty) that do not match the relationship predicted by the hypothesis, or the trend shown by the rest of the data.

Bias – When results are shifted in a particular direction by a systematic error.

Calibrate – Align a measuring instrument's scale with known points or values.

Conclusion – A reasoned interpretation based on experimental data.

Continuous variable – A variable that can take any numerical value within a range.

Controlled variables – Variables not under investigation that are kept constant and monitored during an experiment.

Correlation – When one variable appears to have a relationship with another.

Correlation coefficient – A statistical measure that indicates the degree of relationship between two variables.

Decimal places – The number of digits, including zeros, to the right of the decimal point.

Dependent variable – The variable that is being directly measured in an investigation. Its value depends on the independent variable.

Directly proportional – As one variable increases, the other increases by the same percentage.

Discrete variable – A variable that can only have a certain number of values (that do not have an order), or categories.

Error bars – Graphical representation used on graphs to display the uncertainty (error) in a measurement.

Error propagation – A calculus derived statistical calculation designed to combine uncertainties from multiple variables, in order to provide an accurate measurement of uncertainty.

Evaluation – An assessment of all the limitations of an investigation, the quality of the data and the reliability of the conclusions.

Extrapolation – Estimation of a value for a variable outside the range of the data, by assuming that the relationship between the variables continues to be valid. This is often done using a line graph by extending the line (or curve).

Fair test – A focused experiment, adhering to the scientific method, in which only the independent variable is allowed to significantly affect the dependent variable.

Hazard – The potential to cause harm.

Hypothesis – A proposed explanation based on limited data and observations, the predictions of which may be tested by investigation.

Independent variable – A variable whose value is changed (across a range) by the experimenter in an investigation, to establish its effect on the dependent variable.

Interpolation – Estimation of a value for a variable between two or more known values, or of a scale reading between two scale graduations.

Investigation – A scientific study consisting of a controlled experiment in the laboratory.

Limitations – The restrictions of a particular experimental technique or instrument, which may influence the results.

Line of best fit – A graph line drawn to pass through the plotted points, so that most lie on the line or roughly evenly distributed on either side of the line.

Linearization – Converting a relationship to a linear one by mathematically modifying the variables.

Literature value – An accepted value from the physics literature of a physical constant or experimental measurement.

Mean – An arithmetic average of a set of values. The mean of a set of repeated measurements (with random errors) will give a more accurate result.

Measurement – A record of, or the process of recording, the size of a physical quantity: its numerical amount and its unit.

Methodology – The methods/techniques used to carry out an investigation and their justification.

Outlier – Any value that is numerically distinct from most of the other data points.

Percentage uncertainty (percentage error) – An uncertainty or error in a measured value, expressed as a percentage of the value.

Physical quantity – Property of an object that can be quantified (given a numerical value).

Precision – How close repeated measurements of the same quantity are to their mean value.

Prediction – Expected results of an investigation.

Processed data – Raw experimental data that has been mathematically manipulated.

Propagation of errors – Calculation of the overall uncertainty when processing data containing random errors through a sequence of mathematical operations.

Qualitative data – Observations recorded without measurements being taken.

Quantitative data – Numerical results; measurements with units.

Quantitative relationship – A relationship between variables that can be described by a mathematical equation.

Random error – Experimental error that causes data to vary in an unpredictable way from one measurement to the next. Readings are spread about a mean value.

Range – The difference between the smallest and largest values.

Raw data – Recorded observations and measurements that have not yet been processed or analysed.

Reliability – The extent to which the results of an investigation can be consistently replicated, within limits of experimental uncertainty.

Repeatability – Precision obtained when measurement results are produced in one laboratory, by a single experimenter, using the same equipment under the same conditions.

Replication – A repeating of the entire experiment (on the same occasion with the same apparatus) to record repeat measurements and observations.

Reproducibility – Precision obtained when measurement results are produced by different laboratories (or by different experimenters using different pieces of equipment).

Resolution – The smallest change in reading that can be detected by a measuring instrument.

Risk – The probability of harm occuring.

Risk assessment – A consideration of the hazards that impact human health that could be encountered during an investigation and their level of risk, as well as any environmental impact (for example of disposal).

Scientific method – The use of controlled observations and measurements during an experiment to test a hypothesis.

Scientific notation – A method for expressing a given quantity as a number between 1 and 10, having significant digits necessary for a specified degree of accuracy, multiplied by 10 to the appropriate power.

Secondary data – data obtained from another source, such as via reference material or published third-party results. Analysis of secondary data can form an important part of experimental work.

Sensitivity – The smallest change that can be detected by a measuring instrument.

Significant figures – The digits of a number that are used to express it to the required degree of accuracy, starting from the first non-zero digit.

Simulation – A representation (model) of a process or a system usually involving ICT that imitates a real or an idealized situation.

Standard deviation – A measure of the spread of a set of data from the mean.

Systematic error – Experimental error that causes data to be shifted by a consistent amount each time a measurement is made.

Trend – The general relationship shown by a set of related measurements.

Uncertainty – Defined range of measured values in which the true value is the central point.

Validity (of methodology) – Suitability of the investigative methodology to answer the research question.

Variable – A condition or factor that can vary and may be varied during an investigation and is likely to affect the value of a related quantity.

Zero error – The scale reading of an instrument when the real value is known to be zero. It can be either positive or negative.

Answers

Chapter 1

Page 5

1 Cross-sectional area = 4.9×10^{-8} m²; volume = 9.8×10^{-10} m³

Page 6

2 The equation will take the form $v = km^a l^b T^c$, where the power constants a, b and c are unknown. Re-writing the equation using dimensions:

$[LT^{-1}] = [M^a][L^b][M^c L^c T^{-2c}]$

For L: $1 = b + c$; for M: $0 = a + c$; for T: $-1 = -2c$. Solving gives $a = -\frac{1}{2}$, $b = \frac{1}{2}$ and $c = \frac{1}{2}$.

Hence, $v = km^{-\frac{1}{2}} l^{\frac{1}{2}} T^{\frac{1}{2}}$, where k is an arbitrary constant.

Page 7

3 650 nm, 3 km, 82 kg, 123 kN, 950 MHz, 80 mA

4 0.167 GW = 0.167×10^9 W = 167×10^6 W = 167 MW, which is smaller than 1500 MW.

Page 8

5 a 2.20×10^6 cm³

   b 59 cm²

   c 4.6×10^7 m³

   d 25 m s⁻¹

6 4.243 ly = $4.243 \times 9.46 \times 10^{15}$ m = 4.01×10^{16} m

   4.01×10^{16} m = 4.01×10^{13} km

7 $2 \times 1.60 \times 10^{-19}$ J = 3.20×10^{-19} J

Chapter 2

Page 11

1 a zero error = 0.02 mm; original reading = 6.78 mm; corrected reading = 6.78 − 0.02 = 6.76 mm

   b zero error = 0.03 mm; original reading = 9.60 mm; corrected reading = 9.60 − 0.03 = 9.57 mm

   c zero error = −0.02 mm; original reading = 7.18 mm; corrected reading = 7.18 + 0.02 = 7.20 mm

Page 13

2 $\frac{2.10}{400} \times 10 = 0.0525$ mm

3 a zero error = 0.03 cm; original reading = 7.13 cm; corrected reading = 7.13 − 0.03 = 7.10 cm

   b zero error = −0.04 cm; original reading = 1.97 cm; corrected reading = 1.97 + 0.04 = 2.01 cm

   c zero error = −0.10 cm; original reading = 4.10 cm; corrected reading = 4.10 + 0.10 = 4.20 cm

Page 15

4 $\sin \theta = \frac{85}{1000} = 0.085$; $\theta = 4.876° \simeq 4.9°$

Assuming an uncertainty in the height and length measurements of ±2 mm (that is, ±1 mm in the reading at each end):

% uncertainty in height = $\frac{2}{85} \times 100 = 2.4\%$

% uncertainty in length = $\frac{2}{1000} \times 100 = 0.2\%$

Uncertainty in sine of angle = 2.4% + 0.2% = 2.6%

Page 20

5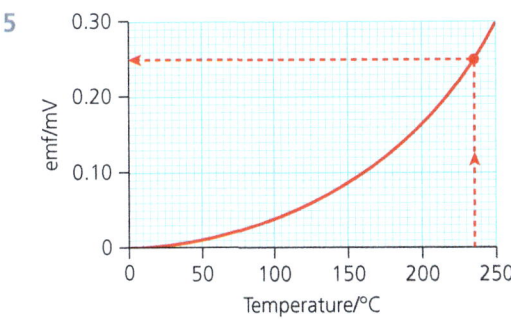

The interpolation line intersects at 235 °C.

Chapter 4

Page 37

1

Height of free fall, h/m ±0.005 m	Average time of free fall, t/s ±0.01 s	t^2/s^2
0.600	0.39	0.15
0.800	0.45	0.20
1.000	0.48	0.23
1.200	0.53	0.28
1.400	0.57	0.32
1.600	0.60	0.36

Relationship: $2s = gt^2$

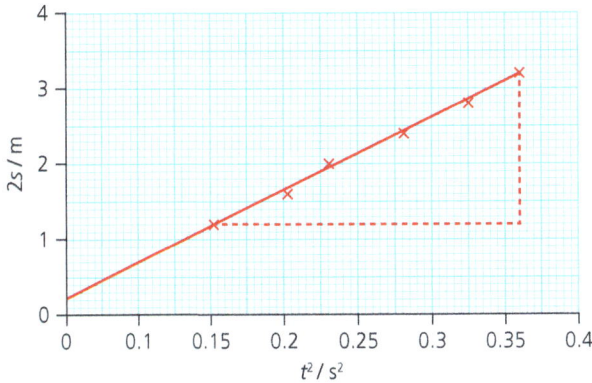

Gradient = g

Therefore, $g = \frac{2}{0.21} = 9.5\,\mathrm{m\,s^{-2}}$

A small negative intercept on the t^2 axis suggests a systematic error, either in t (possibly due to a delay in the timer starting after the ball has been released) or in the measurement of h (all the h values being too large).

Page 41

2 a Heat lost to the surroundings when the system is above room temperature will cancel out the heat taken in from the surroundings when the system is below room temperature.

b Mass of ice = final mass (of calorimeter + water + ice) − initial mass (of calorimeter + water)

c $(mc\,\Delta\theta)_{Al} + (mc\,\Delta\theta)_{water} = (ml)_{ice} + (mc\,\Delta\theta)_{melted\ ice}$

giving $l_{ice} = 3.2 \times 10^5$ J kg^{-1}

d Possible reasons: thermometer not sensitive enough; lack of thermal insulation; lack of stirring to spread heat; heat loss/gain to surroundings; too long for ice to melt; inside of aluminium calorimeter tarnished; splashing; heat capacity of thermometer significant.

Percentage error = [theoretical value − experimental value]/theoretical value × 100 = [3.3 × 10^5 J kg^{-1} − 3.2 × 10^5 J kg^{-1}]/3.3 × 10^5 J kg^{-1} × 100 = 3%

Page 44

3 a, b

Pressure, p/kPa	500	245	170	125	100	50
Volume of gas, V/cm³	1.00	2.00	3.00	4.00	5.00	10.00
pV/kPa cm³	500	490	510	500	500	500
V⁻¹/cm⁻³	1.00	0.50	0.33	0.25	0.20	0.10

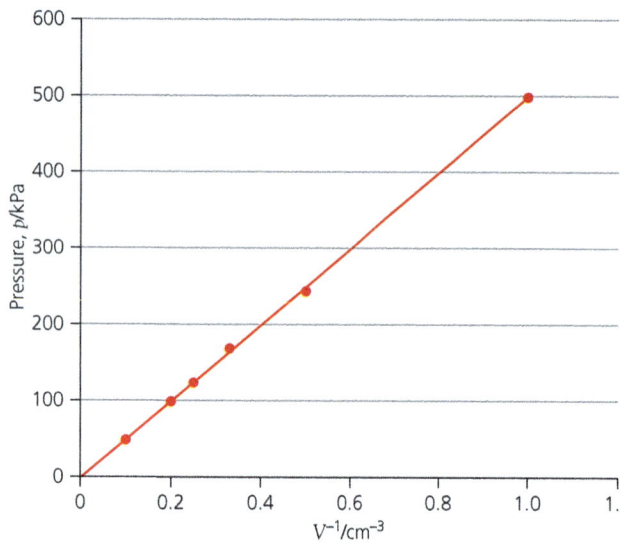

c A graph of log p versus log v (at constant temperature) will give a straight line with a gradient of −1 for ideal results that verify Boyle's law.

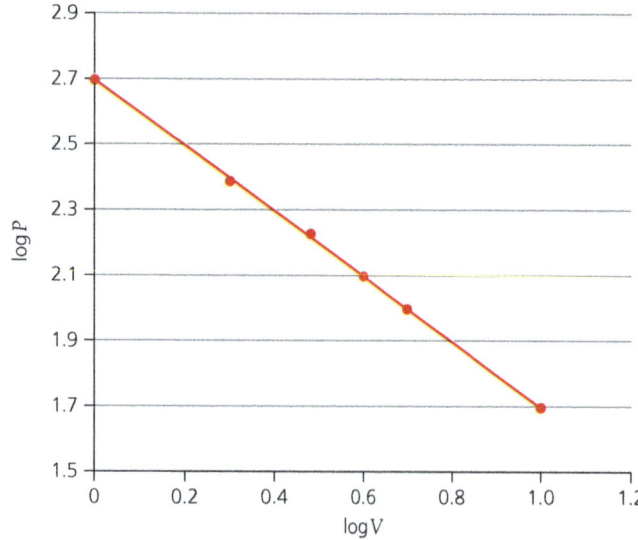

Page 46

4

Temperature/°C	Absolute temperature, T/K	Volume of gas, V/dm³	V/T/dm³ K⁻¹
273	546	0.1094	0.000 20
100	373	0.0748	0.000 20
10	283	0.0568	0.000 20
1	274	0.0545	0.000 20
0	273	0.0544	0.000 20
−20	253	0.0503	0.000 20

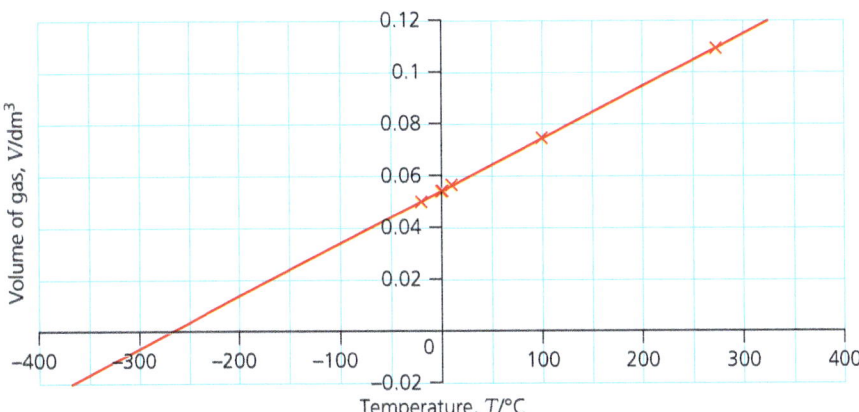

From the graph, absolute zero is approximately −270 °C.

Page 47

5 In the Kundt's tube experiment, a horizontal air column is adjusted until resonance is achieved for a specific frequency of sound. When this happens, the cork dust in the tube is disturbed at the displacement antinodes (A), forming small heaps at the nodes (N). The location of the nodes enables the measurement of the wavelength of the sound.

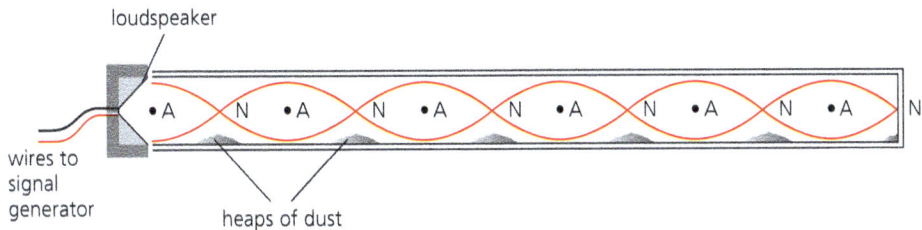

Page 49

6

Angle of incidence, θ_1/°	Angle of refraction, θ_2/°	$\sin \theta_1$	$\sin \theta_2$	$\dfrac{\sin \theta_1}{\sin \theta_2}$
30	19	0.500	0.326	1.56
45	28	0.707	0.469	1.51
65	37	0.906	0.602	1.51

The data show that $\dfrac{\sin \theta_1}{\sin \theta_2}$ is (approximately) constant; this verifies Snell's law.

Page 50

7 a $n = \dfrac{1}{\sin \theta_c} = \dfrac{1}{\sin 42°} = 1.49$

   b $\sin \theta_c = \dfrac{1}{n} = \dfrac{1}{1.46} = 0.685$

      $\theta_c = \sin^{-1} 0.685 = 43.2°$

Page 52

8

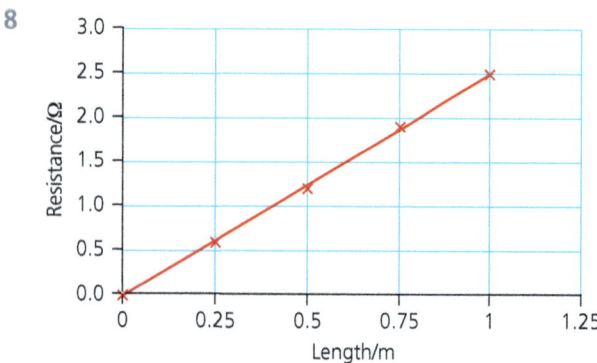

The straight line is a good fit for the data points, indicating that resistance is directly proportional to the length of the wire. This means that longer wires have higher resistance.

Every length of wire will resist electrons to a certain degree. The longer the wire, the longer the path the electrons have to flow; in a longer path, electrons experience more collisions with ions and hence the wire has greater resistance.

9

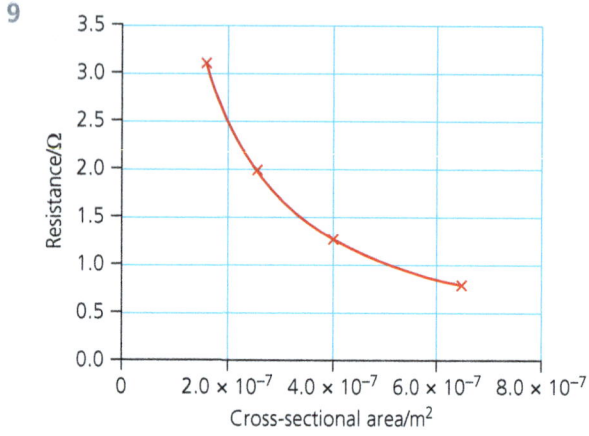

The graph forms a hyperbola. The equation of the line of best fit has a power of −0.9994 which is approximately −1, indicating an inverse relationship between resistance and cross-sectional area. This means that thicker wires have lower resistance.

Thicker wires have a wider and larger path for the electrons to flow through, compared with thinner wires (which tend to constrict or narrow the electron path and hence resist it more due to more collisions).

Page 55

10 a

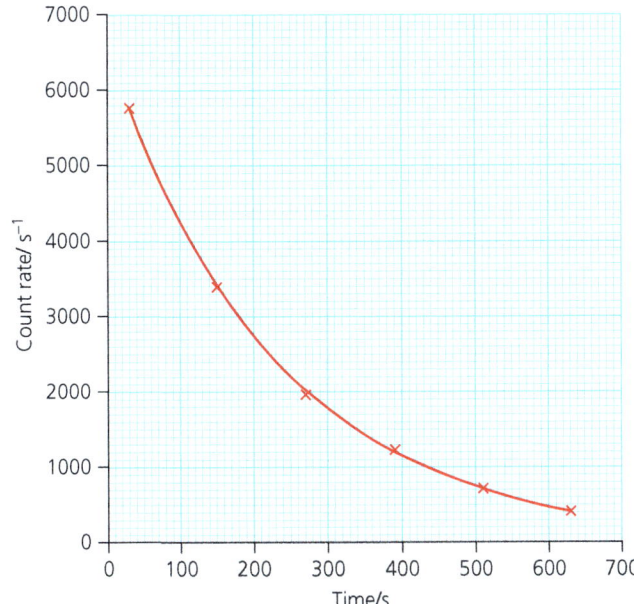

b

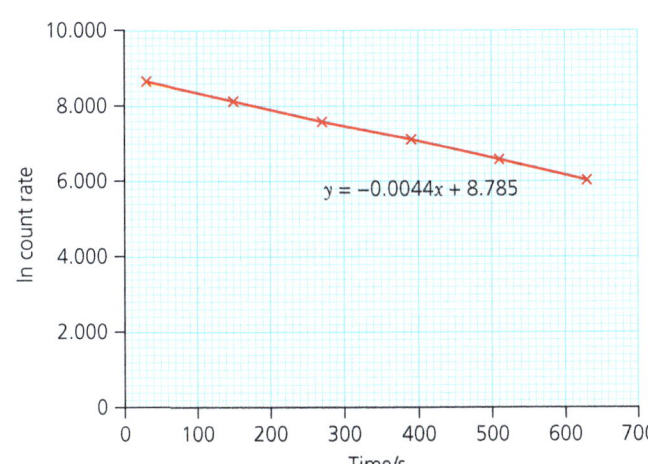

$A = A_0 e^{-\lambda t}$

Therefore, $\ln A = \ln A_0 - \lambda t = -\lambda t + \ln A_0$

This equates to a straight line graph with gradient $-\lambda$ and y-axis intercept $\ln A_0$

Therefore, $\lambda = 0.0044$ and $A_0 = 6535$

Page 57

11 a Fringe separation will decrease.

  b Fringe separation not affected.

  c Fringe separation will decrease.

  d Fringe separation not affected (but the fringes will be brighter).

  e Fringe separation will increase (because the wavelength is greater).

Chapter 5

Page 62

1. a. 1.002×10^3

    5.4×10^1

    6.9263×10^9

    -3.93×10^2

    3.61×10^{-3}

    -3.8×10^{-3}

   b. 1930

    30.52

    −0.0429

    6 261 000

    0.000 000 095 13

    0.000 000 000 095 12

Page 63

2. 2, 4, 2, 4, 2, 5, 1

3. a. 1.237
   b. 1.238
   c. 0.0135
   d. 2.1

Page 64

4. 3.6532 (4 dp); 3.653 (3 dp); 3.65 (2 dp)

Page 70

5. Mean time = 4.8 s

   Uncertainty = ±0.4 s

6. Mean = 0.1015

   Standard deviation = 3.7×10^{-4}

   Data to be discarded if $|Z_{calc}| \geq Z_{0.025}$

   that is if $|Z_{calc}| \geq 1.96$

   $$\frac{0.1015 - 0.1021}{(\sqrt{1.4} \times \frac{10^{-7}}{\sqrt{5}})} = 3.586$$

   Since $|Z_{calc}| \geq 1.96$, discard 0.1021

Page 71

7. $a + b = (5.5 \pm 0.3)$ cm

   $a - b = (0.9 \pm 0.3)$ cm

   $a \times b = (7.4 \pm 0.8)$ cm

   $\frac{b}{a} = (0.7 \pm 0.07)$ cm

8. Radius = (2.3 ± 0.1) cm

   Circumference = $2\pi(2.3 \pm 0.1)$ cm = (14.4513 ± 0.6283) cm = (14 ± 1) cm (to 1 sf)

Page 73

9 Steepest gradient = 347 nF

Best fit gradient = 319 nF

Shallowest gradient = 285 nF

Percentage uncertainty in gradient = 11%

Absolute uncertainty in capacitance = 35 nF

Capacitance = (321 ± 35) nF

Answers within a range of 10% of stated values should be accepted.

Chapter 6

Page 78

1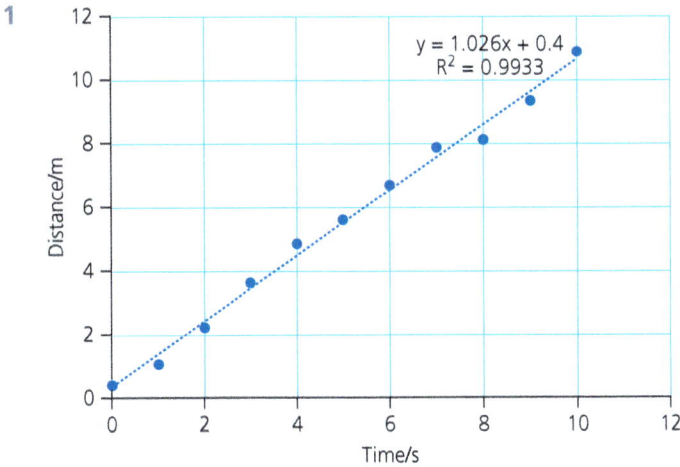

Page 80, page 81

2 & 3 Learners should create a spreadsheet with over 10,000 rows, where all the values are dependent on random numbers. There are no definitive answers; accept spreadsheets as correct if the formulae have been entered correctly.

Page 85

4 Advantages include:
- reaction time eliminated
- automatically records current and time readings simultaneously
- updates the display more rapidly than a standard ammeter
- high sample rate means it is possible to record many pairs of readings
- almost immediately measures and records the initial current, the instant the experiment begins
- the raw data are stored and can be easily processed or presented as a graph and analysed.

5 30 s × 0.50 s^{-1} = 15 measurements

Chapter 8

Page 98

1. Mpemba effect: To find out which freezes faster: a fixed mass of water (de-ionized and free of dissolved air) at 90 °C or the same mass of water at 35 °C, when placed in a freezer in identical glass beakers.

   (See http://qoptics.byu.edu/Physics416/FirstReading.pdf or https://www.nature.com/articles/srep37665)

   Kaye effect: To establish the height at which the Kaye effect is observed for liquid soap when a thin stream of soap solution is poured into a watch glass of the fluid.

   (See https://skullsinthestars.com/2013/03/29/physics-demonstrations-a-short-discussion-of-the-kaye-effect/ or http://www.flowvis.org/wp-content/uploads/2016/11/84487-142124-Joseph-Straccia-Nov-8-2016-208-PM-JStraccia_Team_Second_Report_110916.pdf)

Page 105

2. Research question: To determine the relationship between the terminal velocity of a steel ball bearing falling through olive oil and its radius.

   Independent variable: radius of ball bearing

   Dependent variables: distance fallen; time taken to fall that distance

   Processed dependent variable: terminal velocity

   Controlled variables: density, temperature and viscosity of the olive oil; cross-sectional area of the measuring cylinder; density of the steel of the ball bearing.

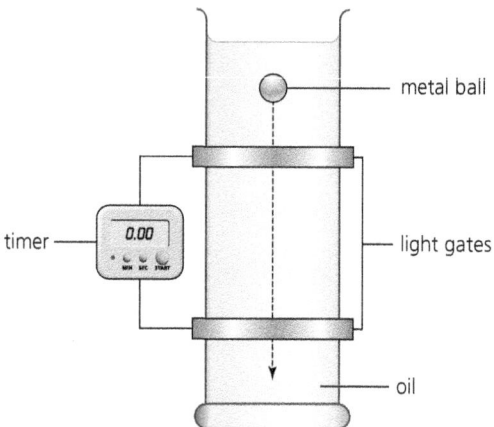

Investigating the terminal velocity of a steel ball bearing in olive oil

A steel ball of known radius (measured by vernier calipers) is released by tweezers from the surface of olive oil (of known temperature measured by a temperature sensor). The time taken to fall a fixed vertical distance (measured using a ruler and a set square) is recorded. This is repeated three times. The experiment is then repeated with five steel balls of the same composition but with a different radius; three times are recorded for each ball. The measuring cylinder must be sufficiently tall that the ball reaches terminal velocity.

The radius of the steel ball bearing is half the diameter. Speed = distance ÷ time. Stoke's law suggests that terminal velocity is proportional to the square of the radius of the steel ball bearing. A larger radius increases the resistive force, resulting in a smaller terminal velocity. A graph of average terminal velocity (y-axis) versus radius (x-axis) of ball bearing should be drawn. A power law curve of best fit should be added.

Safety

Safety glasses must be worn. Wash hands afterwards or as necessary.

In case of skin contact with olive oil, wash with soap and water.

In case of eye contact with olive oil, flush with plenty of water or eye wash solution for 15 minutes.

Clean up any spillages immediately using absorbent material and wash the area with detergent and water. Spilled oil presents a slip hazard.

Keep olive oil away from sources of ignition and naked flames.

Index

A

absolute uncertainty (absolute error) 66
academic honesty 128
acceleration due to gravity, mandatory practical 37–8
accuracy ix, 65
   significant figures 62–4
air tracks 35
alpha decay 54
ammeters 21–4
analogue meters 22
analysis ix
   checklist 112–14
   data processing 108
   data recording 107–8
   error propagation 110–11
   interpretation 111
   mark descriptors 107
   rejecting data 111
   validity of methodology 111, 112
angles 15
angular magnification of a microscope 60
anomalous data x
approaches to learning (ATLs) xiv–xv
area measurements 14–15
assumptions 119

B

background information 98, 105
background radiation 29
background theory 104–5
balances 16
beta decay 54
bias 117
Boyle's law 43–4
burettes 16–17
   uncertainty 67

C

calculations, significant figures 63–4
calibration 19
calibration error 24
calorimeters 42
cathode ray oscilloscope (CRO) 24–5
charging an electroscope 26–7
Charles' law 43–6
circuit symbols 125
coding, Python programming 82–4
communication
   academic honesty 128
   checklist 129
   mark descriptors 123
   referencing 127
   relevance and conciseness 123–4
   report format 125–6
   structure and clarity 123
   terminology and conventions 124–5

compound microscopes, mandatory practical 58–60
computer simulations xiv, 79–84, 87
concept tree xi
conclusions 111, 115–17, 121
contact, charging by 26
continuous variables x
controlled variables ix, 99–100
conventions 124–5, 129
converting units 7
correlation x
correlation coefficients 78
cosine function 76–7
critical angle 50
current balances 29–30
current measurement 21–4

D

data ix, x
   secondary xii, 120
data analysis 78–9
databases 86–7
data collection 106
data interpretation 111, 114
data logging 84–5
data presentation 113
   graphs 109–10
   tables 107
data processing 108, 113
   significant figures 65
data recording 107–8, 112
decimal places 63
dependent errors 69
dependent variables ix, 99–100
depth, real and apparent 50
derived units 5
dimensional analysis 6
dimensionless quantities 4
diode bridge rectification circuits, mandatory practical 57–8
direct proportion 16
discrete variables x

E

electricity, safety considerations 102
electroscopes 26–7
environmental impact 102–3, 105
equations 124
   dimensional analysis 6
error bars 72
error propagation 71, 110–11
error reduction 67–9
errors 9, 66, 108, 113
   human 116
   independent and dependent 69
   oscilloscopes 24
   sources of 70, 117–19
   see also uncertainties

estimating 8
ethics 102, 105
evaluation ix
   checklist 121–2
   conclusions 115–17
   improvements and extensions 119–20
   mark descriptors 115
   of strengths and weaknesses 117–19
   using secondary data 120
Excel *see* spreadsheets
exploration
   background information 98
   checklist 105–6
   ethics 102
   hypothesis construction 99
   mark descriptors 94
   methodology 100–1
   report structure 103–5
   research questions 94–8
   safety considerations 102–3
   variables 99–100
exponential functions 75
exponents, significant figures 64
extensions 119–20, 122
extrapolation 45

F

fair tests 94
free-fall, mandatory practical 37–8
function generators 36
fundamental (base) units 4
fusion, specific latent heat of 40–1

G

galvanometers 22
gamma emission 54
gas laws 43–6
gas pressure measurement 21
Geiger–Müller tubes 28–9
geometric properties 14
gold leaf electroscope 26–7
grade boundaries xvi
gradients 72–4
graphs 72, 109–10, 124
   gradients 72–4
   linearization ix, 75, 76
   types of 74–7
   using a spreadsheet 78
gravity, acceleration due to, mandatory practical 37–8

H

half-life, mandatory practical 53–5
Hall effect 28
Hall probes 27–8
harmonics 47
hazard identification 102–3
hazard warning signs 102

heating, safety considerations 103
heat loss minimisation 42
hot-wire ammeters 23
hypotheses ix, 99, 104

I

IA map 126
IA report
    analysis 107–14
    communication 123–9
    evaluation 115–22
    exploration 94–106
    grade boundaries xvi
    marking criteria xvi
    personal engagement 90–3
    planning xvi–xvii
ice, specific latent heat of fusion 40–1
improvements 119–20, 122
independent errors 69
independent variables ix, 99–100
induction, charging by 27
internal resistance, mandatory practical 52–3
interpolation 15
interpretation 111, 114
investigation cycle viii–ix
investigations 94–7

L

lasers 34–5
learner profile xv
Leidenfrost effect 20
length measurement 9–14
length of report 123
lever balances 16
light gates 18
limitations 101, 117–19, 120
linearization of graphs ix, 75, 76
lines of best fit 72
LINEST function, Excel 73–4
literature values 41
logarithms, significant figures 64
log graphs 75–6

M

magnetic flux density measurement, Hall probes 27–8
magnetism, safety considerations 102
mandatory practicals xii
    acceleration of free-fall due to gravity 37–8
    compound microscopes 58–60
    diode bridge rectification circuit 57–8
    gas laws 43–6
    half-life 53–5
    internal resistance 52–3
    refracting telescopes 60–1
    refractive index 48–51
    resistance 51–2
    resonance 47–8
    specific heat capacities 38–40
    specific latent heats 40–2
    speed of sound 46–8
    Young's double-slit experiment 55–7
manometers 21
mark descriptors

analysis 107
communication 123
evaluation 115
exploration 94
personal engagement 90
marking criteria xvi
mass measurement 16, 67
mean 69
measurements ix
    angles 15
    area 14–15
    current and potential difference 21–4
    error reduction 67–9
    gas pressure 21
    length 9–14
    mass 16
    movement 18–19
    sources of error 70
    temperature 19–20
    time 18
    uncertainty estimation 69
    volume 14–15, 16–17
measuring cylinders 16–17
    uncertainty 67
mechanics, safety considerations 103
method of mixtures 38–9
methodology xvi, 100–1, 106
    report structure 103
    validity 111, 112
metre rules 9, 14
    uncertainty 66–7
metric multipliers 7
micrometer screw gauges 10–11, 14
microscopes, compound 58–60
movement measurement 18–19
multimeters 23–4

N

non-SI units 8
null methods 119

O

optical benches 32–3
oscillation period measurement 67
oscilloscopes 24–5
outliers 70
overflow (Eureka) cans 17

P

parallax error 9
percentage uncertainty (percentage error) 15
personal engagement 90–1
    checklist 93
    evidence of personal input 92–3
    ideas for investigations 91
    justifying research questions 92
    mark descriptors 90
photoelectric effect 27
physical quantities 4
physics
    concept tree xi
    nature of x
pipettes 16
    uncertainty 67
plagiarism 128

planning xvi–xvii
potential difference measurement 21–5
practicals xi–xiv
    suggested xii–xiii
    see also mandatory practicals
precision ix, 63, 65
predictions ix
prefixes 7
pressure law 43–5
pressure measurement, gases 21
processed data ix, x
propagation of errors 71, 110–11
proportions, direct 16
protactinium-234, half-life 54–5
protractors 15
Python programming 82–4

Q

quadratic functions 75
qualitative data x, 108
quantitative relationships ix
quantitative data x

R

radiation detection 28–9
radioactive decay 54
    investigating half-life 53–5
    simulation 80–2
random errors 9, 67–9, 108, 117–18
random number generation 79
range of values 62
raw data ix, x
ray boxes 32
ray diagrams 33
reading error 24
recording data 107–8, 112
rectification, mandatory practical 57–8
referencing 127
refracting telescopes, mandatory practical 60–1
refractive index, mandatory practical 48–51
rejecting data 111
relevance 123, 129
reliability x
repeatability 66
replication x, 119
report structure 103–5
reproducibility 66
research questions 92, 105
    hypothesis construction 99
    phrasing 97–8
    report structure 103
    selection 94–7
resistance, mandatory practical 51–2
resolution 14
results, report structure 104
ripple tanks 30–1
risk assessment xvi, 26, 102–3
    lasers 34
rounding off 63

S

safety considerations 101, 102–3, 105
    lasers 34–5
    radioactive materials 29
    report structure 103

scientific method viii–ix
scientific notation 62
secondary data xii, 120
sensitivity ix
signal generators 36
significant figures 62–5
simple harmonic motion, simulation 83–4
simulations xiv, 79–84, 87
sine function 76–7
SI units 4–5
smartphone apps 86
sonometers 31–2
specific heat capacities, mandatory practical 38–40
specific latent heats, mandatory practicals 40–2
spectrometers 33–4
speed of sound 46–8
spelling 125
spherometer 33
spreadsheets
    data analysis 78–9
    graphs 78
    simulations 79–82
spring balances 16
standard deviation 70
standard masses, uncertainty 67
steam, specific latent heat of vaporization 41–2
stopwatches 18
straight line graphs 74
strengths and weaknesses 117–19, 121

structure and clarity 123, 129
suggested practicals xii–xiii
systematic errors 9, 67–9, 108, 117–18

■ T
tables 107
tangent function 76–7
telescopes, refracting 60–1
temperature measurement 19–20, 42, 67
terminology 124–5, 129
thermocouples 19–20
thermometers 19–20
    uncertainty 67
ticker-tape timers 18–19
time measurement 18
    oscilloscopes 25
total internal reflection 50
travelling microscopes 13–14
trends 85
trigonometric graphs 76–7

■ U
uncertainties x, 66, 113
    combination of 71
    estimation of 69
    in gradients 73–4
    percentage uncertainty (percentage error) 15
    raw data 108
    for specific apparatus 66–7
    see also errors
units 124

    conversion between 7
    derived 5
    fundamental 4
    non-SI 8
    prefixes 7

■ V
validity of methodology 111, 112
van de Graaff generator 27
vaporization, specific latent heat of 41–2
variables ix, x, 99–100
    report structure 103
vernier calipers 12–13, 14
vernier scales 9–10
voltmeters 21–4
volume measurements 14–15, 16–17

■ W
wave investigations 30–2
weight 16

■ Y
Young's double-slit experiment, mandatory practical 55–7

■ Z
zero errors 9